Cover Illustration by Russ Deverniero

The Best Enemy
Money Can Buy

Antony C. Sutton

Foreword by
Gary North, Ph.D.

Liberty House Press — Billings, Montana
1986

**Dedicated to the memory of those who died in
Korea and Vietnam — victims of our
own technology and greed.**

This business of lending blood money is one of the most thoroughly sordid, cold blooded, and criminal that was ever carried on, to any considerable extent, amongst human beings. It is like lending money to slave traders, or to common robbers and pirates, to be repaid out of their plunder. And the man who loans money to governments, so called, for the purpose of enabling the latter to rob, enslave and murder their people, are among the greatest villains that the world has ever seen.

LYSANDER SPOONER, No Treason (Boston, 1870)

TABLE OF CONTENTS

CONCLUSIONS:

Foreword
Gary North

In December of 1979, the Soviet Union launched a lightning-fast military offensive against the backward nation of Afghanistan. It was after this invasion that President Jimmy Carter admitted publicly that it had taught him more about the intentions of the Soviets than everything he had ever learned. Never again would he kiss the cheeks of Premier Brezhnev before the television cameras of the West. The Democrat-controlled Senate even refused to ratify his SALT II treaty. (By the way, President Reagan has been honoring its terms unofficially, and he already has ordered the destruction of several Poseidon submarines, including the **U.S.S. Sam Rayburn**, the dismantling of which began in November of 1985,[1] and which cost a staggering $21 million for the destruction of that one ship.[2] The **Nathan Hale** and the **Andrew Jackson** are scheduled for destruction in 1986.[3] To comply with SALT II, we will have to destroy an additional 2,500 Poseidon submarine warheads. "Good faith," American diplomatic officials argue. ("Good grief," you may be thinking.)

The invasion of Afghanistan was a landmark shift in Soviet military tactics. Departing from half a century of slow, plodding, "smother the enemy with raw power" tactics, the Soviet military leadership adopted the lightning strike. Overnight, the Soviets had captured the Kabul airfield and had surrounded the capital city with tanks.[4]

Tanks? In an overnight invasion? How did 30-ton Soviet tanks roll from the Soviet border to the interior city of Kabul in one day? What about the rugged Afghan terrain?

The answer is simple : there are two highways from the Soviet Union to Kabul, including one which is 647 miles long. Their bridges can support tanks. Do you think that Afghan peasants built these roads for yak-drawn carts? Do you think that Afghan peasants built these roads at all? No, **you** built them.

In 1966, reports on this huge construction project began to appear in obscure U.S. magainzes. The project was completed the following year. It was part of Lyndon Johnson's Great Society. Soviet and U.S. engineers worked side by side, spending U.S. foreign aid money and Soviet money, to get the highways built. One strip of road, 67 miles long north through the Salang Pass to the U.S.S.R., cost $42 million, or $643,000 per mile. John W. Millers, the leader of the United Na-

---

[1]*Washington Times* (Dec. 24, 1985).

[2]Howard Phillips, *Washington Dateline* (Dec. 1985), p. 6.

[3]*Washington Post* (Nov. 27, 1985).

[4]Edward Luttwak, *The Grand Strategy of the Soviet Union* (New York: St. Martin's Press, 1983), ch. 1.

tional survey team in Afghanistan, commented at the time that it was the most expensive bit of road he had ever seen. The Soviets trained and used 8,000 Afghans to build it.[5]

If there were any justice in this world of international foreign aid, the Soviet tanks should have rolled by signs that read: "U.S. Highway Tax Dollars at Work."

Nice guys, the Soviets. They just wanted to help a technologically backward nation. Nice guys, American foreign aid officials. They also just wanted to help a technologically backward nation . . . the Soviet Union.

Seven Decades of Deals

The story you are about to read is true. The names have not been changed, so as not to protect the guilty.

In the mid-1970's, the original version of this book led to the destruction of Antony Sutton's career as a salaried academic researcher with the prestigious (and therefore not quite ideologically tough enough) Hoover Institution on War, Revolution and Peace. That was a high price for Sutton to pay, but not nearly so high as the price you and I are going to be asked to pay because of the activities that this book describes in painstaking detail.

Lenin is supposed to have made the following observation:

"If we were to announce today that we intend to hang all capitalists tomorrow, they would trip over each other trying to sell us the rope."

I don't think he ever said it. However, someone who really understood Lenin, Communism, and capitalist ethics said it. This book shows how accurate an observation it is.

Antony Sutton is not about to offer the following evidence in his own academic self-defense, so I will. Perhaps the best-informed American scholar in the field of Soviet history and overall strategy is Prof. Richard Pipes of Harvard University. In 1984, his chilling book appeared, *Survival Is Not Enough: Soviet Realities and America's Future* (Simon & Schuster). His book tells at least part of the story of the Soviet Union's reliance on Western technology, including the infamous Kama River truck plant, which was built by the Pullman-Swindell company of Pittsburgh, Pennsylvania, a subsidiary of M. W. Kellogg Co. Prof. Pipes remarks that the bulk of the Soviet merchant marine, the largest in the world, was built in foreign shipyards. He even tells the story (related in greater detail in this book) of the Bryant Chucking Grinder Company of Springfield, Vermont, which sold the Soviet Union the ball-bearing

---

[5]"Rugged Afghan Road Jobs Fill Gaps in Trans-Asian Network," *Engineering News-Record* (Nov. 3, 1966).

machines that alone made possible the targeting mechanism of Soviet MIRV'ed ballistic missiles. And in footnote 29 on page 290, he reveals the following:

> In his three-volume detailed account of Soviet purchases of Western equipment and technology . . . [Antony] Sutton comes to conclusions that are uncomfortable for many businessmen and economists. For this reason his work tends to be either dismissed out of hand as "extreme" or, more often, simply ignored.

Prof. Pipes knows how the academic game is played. The game cost Sutton his academic career. But the academic game is very small potatoes compared to the historic "game" of world conquest by the Soviet empire. We are dealing with a messianic State which intends to impose its will on every nation on earth — a goal which Soviet leaders have repeated constantly since they captured Russia in their nearly bloodless **coup** in October of 1917.

Sutton identifies the deaf mute blindmen who sell the Soviets the equipment they need for world conquest. But at least these deaf mute blindmen get something out of it: money. Not "soft currency" Soviet rubles, either; they get U.S. dollars from the Soviets, who in turn get long-term loans that are guaranteed by U.S. taxpayers. Their motivation is fairly easy to understand. But what do the academic drones get out of it? What do they get for their systematic suppression of the historical facts, and their callous treatment in book reviews of works such as Sutton's monumental three-volume set, *Western Technology and Soviet Economic Development*? What was in it, for example, for C. H. Feinstein of Clare College, Cambridge Unversity, who reviewed Sutton's first volume, covering 1917-1930? He could not honestly fault Sutton's basic scholarship, nor did he try:

> . . . he has examined a vast amount of information, much of it previously unknown to scholars, regarding the trading contacts and contracts between the U.S.S.R and the West, notably Germany and the United States. The primary sources were the fascinating and extraordinarily detailed files of the U.S. State Department and the archives of the German Foreign Ministry, and these were supplemented by a wide-ranging and multilingual selection of books and journals.

He even wrote that "Sutton's prodigious researches (and this is apparently only the first of three projected volumes) have provided students of Soviet economic development with a detailed survey of the way in which 'Western' technology was transferred to the Soviet Union, and for this we are indebted to him." But having admitted this — thereby preserving the surface appearance of professional integrity — Feinstein then lowered the academic boom:

Unfortunately, his attempt to go beyond this, and to assess the significance of this transfer and of the concessions policy, is unsatisfactory and overstates the extent and impact of the concessions as well as their importance for Soviet economic development . . . . the defects of Sutton's approach . . . a similar lack of understanding . . . Sutton exaggerates . . . He further indulges his fondness for exaggeration . . . .[6]

You get the basic thrust of the review. "Facts are fine; we are all scholars here." But even the mildest sort of first-stage conclusions concerning the importance and significance of such facts are anathema, for the facts show that the Soviet economy should have this sign over it: "Made in the West." Sutton's subsequent two volumes were never reviewed in this specialized academic journal — the journal, above all other U.S. scholarly journals, in which it would have been most appropriate to include reviews of scholarly books on Soviet economic history. The information blackout had begun, and it was augmented by the publisher's own blackout beginning in 1973, a blackout discussed in this book.

Less than three years after Feinstein's review was published, Bryant Chucking Grinder Co. sold the Soviets the ball-bearing grinders that subsequently placed the West at the mercy of the Soviet tyrants. At last, they possessed the technology which makes possible a relatively low-risk first-strike by Soviet missiles against our missiles and "defenses."[7] Until Bryant supplied the technology, the Soviets couldn't build such offensive weapons, which is why they had lobbied from 1961 until 1972 to get the U.S. government's authorization to buy the units. Within a few years after delivery, they had the missiles installed. Then they invaded Afghanistan. So much for Sutton's "exaggerations."

This book is not really designed to be read word for word. It is a kind of lawyer's brief, filled with facts that none of us will remember in detail. But if the facts were not included, the book's thesis would be too far-fetched to accept. He therefore includes pages and pages of dull, dreary details — details that lead to an inescapable conclusion: that the West has been betrayed by its major corporate leaders, with the full compliance of its national political leaders.

[6]Review of Antony Sutton, *Western Technology and Soviet Economic Development, 1917-1930* (Stanford: Hoover Institution on War, Revolution and Peace, Stanford University, 1968), in *The Journal of Economic History*, XXIX (December 1969), pp. 816-18.

[7]Actually, the United States has no defenses. What we have is an arsenal of retaliatory offensive weapons aimed at Soviet cities, not at Soviet military targets. This is the infamous strategy of Mutually Assured Destruction (MAD) which was implemented by former Secretary of Defense (!!!) Robert Strange McNamara. If Soviet missiles were to take out the bulk of our land-based missiles in a first strike, we would have little choice but to surrender, since our submarine-launched missiles are too weak and too inaccurate to destroy hardened Soviet missile silos, and the Soviets could threaten a second wave of missiles against our cities if we were to attempt to retaliate. On our present position of military inferiority, see Quentin Crommelin, Jr. and David S. Sullivan, *Soviet Military Supremacy* (Washington, D.C.: The Citizens' Foundation, 1985). This book was a project of UCLA's Defense and Strategic Studies Program.

From this time forward, you can say in confidence to anyone: "The United States financed the economic and military development of the Soviet Union. Without this aid, financed by U.S. taxpayers, there would be no significant Soviet military threat, for there would be no Soviet economy to support the Soviet military machine, let alone sophisticated military equipment." Should your listener scoff, you need only to hand him a copy of this book. **It will stuff his mouth with footnotes.**

It probably will not change the scoffer's mind, however. Minds are seldom changed with facts, certainly not college-trained minds. Facts did not change Prof. Feinstein's mind, after all. The book will only shut up the scoffer when in your presence. But even that is worth a lot these days.

From this day forward, you should never take seriously any State Department official (and certainly not the Secretary of State) who announces to the press that this nation is now, and has always been, engaged in a worldwide struggle against Communism and Soviet aggression. Once in a while, Secretaries of State feel pressured to give such speeches. They are nonsense. They are puffery for the folks out in middle America.

You may note for future reference my observation that Secretaries of Commerce never feel this pressure to make anti-Communist speeches. They, unlike Secretaries of State, speak directly for American corporate interests. They know where their bread is buttered, and more important, who controls the knife.

When it comes to trading with the enemy, multinational corporate leaders act in terms of the political philosophy of the legendary George Washington Plunkett of Tammany Hall: "I seen my opportunities, and I took 'em." Plunkett was defending "honest graft"; our modern grafters have raised the stakes considerably. They are talking about bi-partisan treason.

Author's Preface

Back in 1973 this author published *National Suicide: Military Aid to the Soviet Union*, itself a sequel to a three volume academic study, *Western Technology and Soviet Economic Development*, published by the Hoover Institution, Stanford University. These four books are detailed verifiable catalogs of Western technology used and in use by the Soviet Union, acquired by gift, purchase, illegal diversion or theft.

Taken together, these four volumes constitute an extraordinary commentary on a basic weakness in the Soviet system and an equally extraordinary weakness in Western policy making. The Soviets are heavily dependent on Western technology and innovation not only in their civilian industries, but also in their military programs.

Technology is, of course, the life blood of modern economic development: technology is the difference between the Third World and the advanced 21st century development epitomised by Silicon Valley in California.

Regrettably, most economists are not qualified to explore the role of technology in economic development. Technology is assumed as a "given," whereas it is in fact a dynamic factor, the most dynamic factor many would argue, in modern economic development. Similarly, State Department planners, essentially political scientists, are not at home with technology — sufficiently so that in 1963 State issued papers to the effect that the Soviets had only self-generated technology. Even today State and Commerce appear barely conversant with the extent of Soviet "reverse engineering." Fortunately, Department of Defense is more attuned to technology and among all government departments is **alone** aware of the magnitude of the problem to be described in *The Best Enemy Money Can Buy.*

The deaf mute blindmen — to quote from Lenin — are those multinational businessmen who see no further than the bottom line of the current contract. Unfortunately, these internationalist operators have disproportionate influence in Washington. Consequently, arguments based on the flimsiest of evidence and the most absurd logic that fly in the face of all we know about the Soviets are able not only to be heard in Washington, but even form the basis of our policy.

An inevitable conclusion from the evidence in this book is that we have totally ignored a policy that would enable us to neutralize Soviet global ambitions while simultaneously reducing the defense budget and the tax load on American citizens. Whether we like it or not, technology is a political tool in today's world. And if we want to survive in the face of Soviet ambitions, we will have to use this weapon sooner or later. At the moment the combined efforts of the deaf mute blindmen have been successful. Only an informed, aroused electorate has sufficient potential power to counter their suicidal ambitions.

Antony C. Sutton,
California, January 1986.

CHAPTER I

America's Deaf Mute Blindmen

"To attribute to others the identical sentiments that guide oneself is never to understand others."   Gustav Le Bon

Over the past several decades, quietly, without media attention, many Americans in diverse fields of activity have been pressured into silence, and failing silence, have been removed from their positions or excommunicated from a chosen profession. These men range from historians in Department of State, top level officials in Department of Commerce, engineers working for IBM, to academics in America's leading universities.

In each case threats and pressures which led to censorship, firing, and excommunication track back to the deaf mute blindmen, and their associates in political Washington.

Who are the deaf mute blindmen?

The Russian revolutionary Vladimir Ilych (Ulyanov) Lenin coined the phrase, and whatever we think of Lenin's revolutionary philosophy, we cannot deny his genius in the analysis of capitalists and their motivations. Here is Lenin on the "deaf mute blindmen."

> The Capitalists of the world and their governments, in pursuit of conquest of the Soviet market, will close their eyes to the indicated higher reality and thus will turn into **deaf mute blindmen.** They will extend credits, which will strengthen for us the Communist Party in their countries and giving us the materials and technology we lack, they will restore our military industry, indispensable for our future victorious attacks **on our suppliers.** In other words, they will labor for the preparation for their own suicide.[1]

The Suppressed Higher Reality

What is this "higher reality" that Lenin identifies? It is simply that the Soviet system cannot generate sufficient innovation and technology to become a world power, yet Soviet global ambitions demand that its socialist system challenge and surpass the capitalist systems of the West. Lenin deduced just before he died in 1923 that Soviet communism had an Achilles heel, a fatal defect. In a remarkable about-face Lenin then introduced the New Economic Policy, a return to limited free enterprise and a prelude to a long-lasting cooperation with Western capitalists — the deaf mute blindmen. This policy was repeated by Communist China in the early 1980s.

---

[1]Quoted in Joseph Finder, *Red Carpet* (Holt, Rinehart and Winston, New York, 1984), p. 8

It is knowledge of this "higher reality" that has been ruthlessly suppressed by successive Administrations under political pressure from internationalist businessmen. The State Department, for example, has a disgraceful record of attempting to black out information and present a false picture of historical events. Under John Foster Dulles, Dr. G. Bernard Noble, a Rhodes scholar and an enemy of any attempt to change the Establishment's party-line, was promoted to take charge of the Historical Office at State Department. Two historians, Dr. Donald Dozer and Dr. Bryton Barron, who protested the official policy of distorting information and suppressing historical documents, were railroaded out of the State Department. Dr. Barron, in his book, *Inside the State Department*,[2] specifically charged the State Department with responsibility for the exportation of military technology to the Soviet Union, and listed four examples of highly strategic tools whose export to the USSR was urged by officials of the State Department.

1. Boring mills for manufacture of tanks, artillery, aircraft, and the atomic reactors used in submarines.
2. Vertical boring mills for manufacture of jet engines.
3. Dynamic balance machines used to balance shafts on engines for jet airplanes and guided missiles.
4. External cylindrical grinding machines which a Defense Department expert testified were essential in making engine parts, guided missiles, and radar.

Bryton Barron concludes:

It should be evident that we cannot trust the present personnel of the Department to apply our agreements in the nation's interest any more than we can trust it to give us the full facts about our treaties and other international commitments.

Breathtakingly inaccurate are the only words that can describe State Department claims regarding our military assistance to the Soviet Union. The general State Department line is that the Soviets have a self-developed technology, that trade is always peaceful, that we have controls on the export of strategic goods, and that there is no conceivable relationship between our export to the Soviet Union and Soviet armaments production.

An example will make the point. Here is a statement by Ambassador Trezise to the Senate:

Ambassador Trezise: We, I think, are sometimes guilty, Senator, of a degree of false and unwarranted pride in our industrial and technological might, a kind of arrogance, if you will . . . we are ahead of the Soviet Union in many areas of industry and technology. But a nation that can accomplish the scientific

[2]Bryton Barron, *Inside the State Department* (New York: Comet Press, 1956).

and technological feats the Soviet Union has accomplished in recent years is clearly not a primitive, mudhut economy . . . . It is a big, vigorous, strong, and highly capable national entity, and its performance in the space field and in other fields has given us every indication that Soviet engineers, technicians, scientists, are in the forefront of the scientists, engineers, technicians of the world.

Senator Muskie: So that the urge towards increased trade with Eastern European countries has not resulted in a weakening of the restrictions related to strategic goods?

Ambassador Trezise: I think that is an accurate statement, Senator.

Now we have, we think, quite an effective system of handling items which are in the military area or so closely related thereto that they become strategic items by everybody's agreement.

In fact, at the very time Trezise was making the above soothing statement, critical shipments of strategic materials and equipment were going forward to the Soviet Union. The so-called Export Control laws were a leaky sieve due to outright inefficiency in Departments of State and Commerce.

Censorship has enabled politically appointed officials and the permanent Washington bureaucracy to make such unbelievably inaccurate statements without fear of challenge in Congress or by the American public.

The State Department files are crammed with information concerning U.S. technical and economic assistance to the Soviet Union. The author of this book required three substantial volumes (see Bibliography) just to summarize this assistance for the years 1917-1970. Yet former Secretary of State Dean Rusk, presumably acting on the advice of State Department researchers, stated in 1961, "It would seem clear that the Soviet Union derives only the most marginal help in its economic development from the amount of U.S. goods it receives." A statement flatly contradictory to the massive evidence available in departmental files.

In 1968 Nicholas de B. Katzenbach, Assistant Secretary of State, made a statement that was similarly inconsistent with observable fact, and displayed a fundamental lack of common-sense reasoning:

We should have no illusions. If we do not sell peaceful goods to the nations of Eastern Europe, others will. If we erect barriers to our trade with Eastern Europe, we will lose the trade and Eastern Europe will buy elsewhere. But we will not make any easier our task of stopping aggression in Vietnam nor in building security for the United States.[3]

In fact, aggression in South Vietnam would have been impossible without U.S. assistance to the Soviet Union. Much of the key "European" technology cited derives from U.S. subsidiaries.

Jack N. Behrman, former Deputy Assistant Secretary for International Affairs at the Department of Commerce, repeated the same theme on behalf of the Commerce Department:

> This is the old problem of economic dependency. However, I do not believe that Russia would in fact permit herself to become dependent upon imported sources of strategic goods. Rather she would import amounts additional to her strategic needs, thereby relieving the pressure on her economy by not risking dependence.[4]

In fact, Jack Behrman to the contrary notwithstanding, Soviet Russia is the most dependent large nation in modern history, for wheat as well as technology.

Here's another statement from former Secretary of Commerce Maurice H. Stans.

> **Q:** Is there danger of this country's helping the Russians build a war potential that might be turned against the interests of the free world?
>
> **A:** Under the circumstances, we might be very foolish not to accept business which could create jobs in the United States, when refusing to sell to the Soviet Union would in no way deter their progress.[5]

Suppression of Information

Information suppression concerning Soviet relations with the United States may be found in all administrations, Democrat and Republican, from President Wilson to President Reagan. For example, on November 28, 1917, just a few weeks after the Petrograd and Moscow Bolsheviks had overthrown the democratic and constitutional government of Russia, "Colonel" House (then in Paris) intervened on behalf of the Bolsheviks and cabled President Wilson and the Secretary of State in the "Special Green" cipher of the State Department as follows:

> There has been cabled over and published here [Paris] statements made by American papers to the effect that Russia should be treated as an enemy. It is exceedingly important that such criticisms should be suppressed . . .[6]

[3]House of Representatives, *To Amend the Export-Import Bank Act of 1945* (Washington, DC, 1968), p. 64.

[4]Ibid.

[5]*U.S. News & World Report*, December 20, 1971.

[6]See Antony Sutton, *Wall Street and the Bolshevik Revolution* (New York: Arlington House, 1974).

Suppression of information critical of the Soviet Union and our military assistance to the Soviets may be traced in the State Department files from this 1917 House cable down to the present day, when export licenses issued for admittedly military equipment exports to the USSR are not available for public information. In fact, Soviet sources must be used to trace the impact of some American technology on Soviet military development. The Soviet Register of Shipping, for example, publishes the technical specifications of main engines in Russian vessels (including country of manufacture): this information is not available from U.S. official sources. In November 1971, *Krasnaya Zvezda* published an article with specific reference to the contribution of the basic Soviet industrial structure to the Soviet military power — a contribution that representatives of the U.S. Executive Branch have explicitly denied to the public and to Congress.

Even today U.S. assistance to the Soviet military-industrial complex and its weapons systems cannot be documented from open U.S. sources alone because export license memoranda are classified data. Unless the technical nature of our shipments to the USSR is known, it is impossible to determine their contribution to the Soviet military complex. The national security argument is not acceptable as a defense for classification because the Soviets know what they are buying. So does the United States government. So do U.S. firms. So do the deaf mute blindmen. The group left out in the cold is the American taxpayer-voter.

From time to time bills have been introduced in Congress to make export-license information freely available. These bills have never received Administration support. Nonavailability of current information means that decisions affecting all Americans are made by a relatively few government officials without impartial outside scrutiny, and under political pressure from internationlist businessmen. In many cases these decisions would not be sustained if subjected to public examination and criticism. It is argued by policy-makers that decisions affecting national security and international relations cannot be made in a goldfish bowl. The obvious answer to this is the history of the past seventy years: we have had one catastrophic international problem after another — and in the light of history, the outcome would have been far less costly if the decisions **had** been made in a goldfish bowl.

For instance, little more than a decade after House's appeal to Wilson, Senator Smoot inquired of the State Department about the possible military end-uses of an aluminum powder plant to be erected in the Soviet Union by W. Hahn, an American engineer. State Department files contain a recently declassified document which states why no reply was ever given to Senator Smoot:

No reply was made to Senator Smoot by the Department as the

Secretary did not desire to indicate that the Department had no objection to the rendering by Mr. Hahn of technical assistance to the Soviet authorities in the production of aluminum powder, in view of the possibility of its use as war material, and preferred to take no position at the time in regard to the matter.[7]

Congressional action in the Freedom on Information Act and administrative claims of speedy declassification have not changed this basic situation. Major significant documents covering the history of the past seventy years are buried, and they will remain buried until an outraged public opinion puts some pressure on Congress.

Congress has on the other hand investigated and subsequently published several reports on the export of strategic materials to the Soviet Union. One such instance, called "a life and death matter" by Congress, concerned the proposed shipment of ball bearing machines to the USSR.[8] The Bryant Chucking Grinder Company accepted a Soviet order for thirty-five Centalign-B machines for processing miniature ball bearings. All such precision ball bearings in the United States, used by the Department of Defense for missile guidance systems, were processed on seventy-two Bryant Centalign Model-B machines.

In 1961 the Department of Commerce **approved** export of thirty-five such machines to the USSR, which would have given the Soviets capability about equal to 50 percent of the U.S. capability.

The Soviets had no equipment for such mass production processing, and neither the USSR nor any European manufacturer could manufacture such equipment. A Department of Commerce statement that there were other manufacturers was shown to be inaccurate. Commerce proposed to give the Soviet Union an ability to use its higher-thrust rockets with much greater accuracy and so pull ahead of the United States. Subsequently, a congressional investigation yielded accurate information not otherwise available to independent nongovernment researchers and the general public.

Congressional investigations have also unearthed extraordinary "errors" of judgment by high officials. For example, in 1961 a dispute arose in U.S. government circles over the "Transfermatic Case" — a proposal to ship to the USSR two transfer lines (with a total value of $4.3 million) for the production of truck engines.

In a statement dated February 23, 1961, the Department of Defense went on record **against** shipment of the transfer lines on the grounds

---

[7]U.S. State Dept. Decimal File, 861.659-Du Pont de Nemours & Co/5.

[8]U.S. Senate, Committee on the Judiciary, *Proposed Shipment of Ball Bearing Machines to the U.S.S.R.* (Washington, 1961).

that "the technology contained in these Transfermatic machines produced in the United States is the most advanced in the world," and that "so far as this department knows, the USSR has not installed this type of machinery. The receipt of this equipment by the USSR will contribute to the Soviet military and economic warfare potential." This argument was arbitrarily overturned by the incoming Secretary of Defense Robert McNamara. Secretary McNamara did not allow for the known fact that most Soviet military trucks came from two American-built plants even then receiving equipment from the United States. The Transfermatic machines approved by McNamara had clear and obvious military uses — as the Department of Defense had previously argued. Yet McNamara allowed them to go forward.

Yet another calculated deception of the American public can be traced to the Johnson Administration. In 1966 the U.S. Department of State produced a beautiful, extravagantly illustrated brochure of American hand tools. This was printed in Russian, for distribution in Russia, with a preface — in Russian — by Lyndon Johnson. Requests to the State Departament for a copy of this brochure went unanswered. The book is not listed in official catalogues of government publications. It is not available or even known to the general public. No printer's name appears on the back cover. The publisher is not listed. The author obtained a copy from Russia. Here is the preface:

Hand Tools — USA[9]

Welcome to the "Hand Tools — USA" exhibit — the eighth consecutive exhibit arranged for citizens of the Soviet Union.

At this exhibit you will see samples of various hand tools currently manufactured in the United States — tools that facilitate manual work and make it possible to produce better-quality industrial goods at a much lower cost.

Since the very early days of the history of our country, Americans of all ages have worked with hand tools. In industry and at home, in factories and on farms, in workshops and schools, the hand tool has become indispensable in our lives.

Some of these tools have retained their original simplicity of design; others have acquired entirely new forms and are now used to perform new functions.

We sincerely hope that this exhibit will lead to a better understanding of the American people and their way of life.

/s/ Lyndon B. Johnson

---

[9]Author's translation from Russian of brochure for "Hand Tools — USA" exhibit.

Why all the secrecy? Imagine the public reaction in 1966, when the Soviets were supplying the North Viets with weapons to kill Americans (over 5,000 were killed that year), if it had become known that the State Department had published lavish booklets in Russian for free distribution in Russia at taxpayers' expense.

However, the point at issue is not the wisdom of publication, but the wisdom of concealment. The public is not told because the public might protest. In other words, the public cannot be trusted to see things in the same light as the policymakers, and the policymakers are unwilling to defend their positions.

Further, what would have been the domestic political consequences if it had been known that a U.S. President had signed a document in Russian, lavishly produced at the taxpayers' expense for free distribution in Russia, while Russian weapons were killing Americans in Vietnam with assistance from our own deaf mute blindmen? The citizen-taxpayer does not share the expensive illusions of the Washington elite. The political reaction by the taxpayer, and his few supporters in Congress, would have been harsh and very much to the point.

The Deaf Mute Blindmen

The key party interested in concealment of information about our export to the Soviet Union is, of course, the American firms and individuals prominently associated with such exports, i.e., the deaf mute blindmen themselves.

In general, the American public has a basic right to know what is being shipped and who is shipping it, if the Soviets are using the material against us. The public also has a right to know about the personal interests of presidential appointees and previous employment with firms prominent in trade with the USSR.

Until recently, the firms involved could publicly claim ignorance of the use to which the Soviets put imported Western technology. It is not a good claim, but it was made. From the 1970's on, ignorance of end-use is not a valid claim. The evidence is clear, overwhelming, and readily available: the Soviets have used American technology to kill Americans and their allies.

The claim that publication of license information would give undue advantage to competitors is not the kind of argument that an honest businessman would make. It is only necessary to publish certain basic elementary information: date, name of firm, amount, destination in the USSR, and a brief statement of the technical aspects. Every industry has a "grapevine" and potential business in an industry is always common knowledge.

In any event, suppose there **was** adverse comment about a particular sale to the Soviets? Is this a bad thing? If our policies are indeed viable, why fear public opinion? Or are certain sectors of our society to be immune from public criticism?

Soviet dependency on our technology, and their use of this technology for military purposes, could have been known to Congress on a continuing basis in the 1950s and 1960s if export license information had been freely available. The problem was suspected, but the compilation of the proof had to wait several decades until the evidence became available from Soviet sources. In the meantime, Administration and business spokesmen were able to make absurd statements to Congress without fear of challenge. **In general, only those who had already made up their minds that Soviet trade was desirable had access to license information.** These were the deaf mute blindmen only able to see their own conception of events and blind to the fact that we had contributed to construction of Soviet military power.

In 1968, for example, the Gleason Company of Rochester, New York shipped equipment to the Gorki automobile plant in Russia, a plant previously built by the Ford Motor Company. The information about shipment did not come from the censored licenses but from foreign press sources. Knowledge of license application for any equipment to be used to Gorki would have elicited vigorous protests to Congress. Why? Because the Gorki plant produces a wide range of military vehicles and equipment. Many of the trucks used on the Ho Chi Minh trail were GAZ vehicles from Gorki. The rocket-launchers used against Israel are mounted on GAZ-69 chassis made at Gorki. They have Ford-type engines made at Gorki.

Thus, a screen of censorship vigorously supported by multinational businessmen has withheld knowledge of a secret shift in direction of U.S. foreign policy. This shift can be summarized as follows:

1. Our long-run technical assistance to the Soviet Union has built a first-order military threat to our very existence.
2. Our lengthy history of technical assistance to the Soviet military structure was known to successive administrations, but has only recently (1982) been admitted to Congress or to the American public.
3. Current military assistance is also known, but is admitted only on a case-by-case basis when information to formulate a question can be obtained from nongovernment sources.
4. As a general rule, detailed data on export licenses, which are required to establish the **continuing** and long-run dependence of the Soviet military-industrial complex on the United States, have been made available to Congress only by special request, and have been denied completely to the American public at large.

In brief, all presidential administrations, from that of Woodrow Wilson to that of Ronald Reagan, have followed a bipartisan foreign policy of building up the Soviet Union. This policy is censored. It is a policy of suicide.

Persistent pressure from nongovernmental researchers and knowledgeable individuals has today forced the Administration to at least publicly acknowledge the nature of the problem but still do very little about it. For instance, in an interview on March 8, 1982, William Casey, Director of the Central Intelligence Agency, made the following revealing statement:

We have determined that the Soviet strategic advances depend on Western technology to a far greater degree than anybody ever dreamed of. It just doesn't make any sense for us to spend additional billions of dollars to protect ourselves against the capabilities that the Soviets have developed largely by virtue of having pretty much of a free ride on our research and development.

They use every method you can imagine — purchase, legal and illegal; theft; bribery; espionage; scientific exchange; study of trade press, and invoking the Freedom of Information Act — to get this information.

We found that scientific exchange is a big hole. We send scholars or young people to the Soviet Union to study Pushkin poetry; they send a 45-year-old man out of their KGB or defense establishment to exactly the schools and the professors who are working on sensitive technologies.

The KGB has developed a large, independent, specialized organization which does nothing but work on getting access to Western science and technology. They have been recruiting about 100 young scientists and engineers a year for the last 15 years. They roam the world looking for technology to pick up.

Back in Moscow there are 400 or 500 assessing what they might need and where they might get it — doing their targeting and then assessing what they get. It's a very sophisticated and farflung operation.[10]

Unfortunately, Mr. Casey, who pleads surprise at the discovery, is still concealing the whole story. This author (not alone) made this known to Department of Defense **over 15 years ago,** with a request for information to develop the full nature of the problem. This exchange of letters is reproduced as Appendix A. Nothing was done in 1971. In the past 15 years there has been a superficial change — the Reagan Ad-

[10]United States Senate, *Transfer of United States High Technology to the Soviet Union and Soviet Bloc Nations* Hearings before the Permanent Subcommittee on Investigations, 97th Congress Second Session, May 1982, Washington, D.C., p. 55.

ministration is now willing to admit the **existence** of the problem. It has not yet been willing to face the policy challenge. Until the deaf mute blindmen are neutralized, our assistance for Soviet strategic advances will continue.

CHAPTER II

American Trucks in Korea and Vietnam — For the Other Side

If we do not develop our automobile industry, we are threatened with the heaviest losses, if not defeats, in a future war.

Pravda, July 20, 1927

At the end of World War II the U.S. government appointed an interagency committee to consider the future of the German automobile industry and its war-making potential. This committee concluded that **any** motor vehicle industry in **any** country is an important factor in that country's war potential.

More than half U.S. tanks, almost all armored and half-track vehicles and one-third of guns over 33 millimeter were manufactured in U.S. civilian motor vehicle plants.

Consequently, the committee unanimously recommended:

1. **Any** vehicle industry is a major force for war.
2. German automotive manufacturing should be prohibited because it was a war industry.
3. Numerous military products can be made by the automobile industry, including aerial torpedoes, aircraft cannon, aircraft instruments, aircraft engines, aircraft engines parts, aircraft ignition testers, aircraft machine guns, aircraft propeller subassemblies, aircraft propellers, aircraft servicing and testing equipment, aircraft struts, airframes, and so on. A total of 300 items of military equipment was listed.

A comparison of the recommendations from this committee with subsequent administrative recommendations and policies for the export of automobile-manufacturing plants to the Soviet Union demonstrates extraordinary inconsistencies. **If automobile-manufacturing capacity has "warlike" potential for Germany and the United States, then it also has "warlike" potential for the Soviet Union.** But the recommendations for post-war Germany and the Soviet Union are **totally** divergent. Some of the same Washington bureaucrats (for example, Charles R. Weaver of the Department of Commerce) participated in making **both** decisions.

In brief, any automobile or tractor plant can be used to produce tanks, armored cars, military trucks, other military vehicles and equip-

*The report is *Study by Interagency Committee on the Treatment of the German Automotive Industry from the Standpoint of National Security* (Washington, D.C.: Foreign Economic Administration, July 14, 1945), Report T.I.D.C. No. 12.

ment. A major conclusion reached by a U.S. interagency committee formed to study the war-making potential of the U.S. and German automotive industries was that a motor vehicle industry has enormous military potential.

"The Committee recognized without dissent that [Germany's] motor vehicle industry was an important factor in her waging of war during World War II."

On the basis of its findings, the committee recommended that the manufacture of complete automobiles in Germany be prohibited, that the manufacture of certain parts and subassemblies be "specifically prohibited," and that Germany "should not be permitted to retain in her possession any types of vehicles or particular military application, such as track-laying vehicles, multi-axle vehicles, etc."

The committee further listed more than 300 "war products manufactured by the automotive industry."

These conclusions have been ignored for the Soviet automobile industry, even while the Soviets themselves officially stated their intention to use foreign automobile technology for military vehicles **as early as 1927.** V. V. Ossinsky, a top planner, wrote a series of articles for *Pravda* (July 20, 21 and 22, 1927) with the following warning:

> If in a future war we use the Russian peasant cart against the American or European automobile, the result to say the least will be disproportionately heavy losses, the inevitable consequences of technical weakness. This is certainly not industrialized defense.

The Soviet military-civilian vehicle manufacturing industry, as subsequently developed, produces a limited range of utilitarian trucks and automobiles in a few large plants designed, built by, and almost entirely equipped with Western, primarily American, technical assistance and machinery. These motor vehicle plants mostly manufacture their own components and ship these to assembly plants elsewhere in the Soviet Union.

There is a high degree of integration between Russian military and civilian vehicle models. Military and civilian vehicles have interchangeable parts and Soviet policy is to maximize unification of military and civilian designs to assist model change-over in case of war.

This unification of military and civilian automobile design has been described by the Soviet economist A. N. Lagovskiy:

> The fewer design changes between the old and the new type of product, the easier and more rapidly the enterprise will shift to new production. If, for example, chassis, motors, and other parts of a motor vehicle of a civilian model are used for a military motor vehicle, or course, the shift to the mass production of the military motor vehicle will occur considerably faster and more easily than if the design of all the main parts were different.

Lagovskiy notes that Soviet "civilian" agricultural tractors and motor vehicles can be used directly as military vehicles without major conversion. Soviet tractors (direct copies of Caterpillar models) were used as artillery tractors in World War II and Korea. General G. I. Pokrovski makes a similar argument about the U.S. 106-millimeter recoilless weapon mounted on a Willys jeep and comments that "even relatively powerful recoilless artillery systems can, at the present time, be mounted on a light automobile without reducing the number of men accomodated."[11]

Almost all — possibly 95 percent — of Soviet military vehicles are produced in very large plants designed by American engineers in the 1930s through the 1970s.

The Soviet Military Truck Industry

Soviet civilian and military trucks are produced in the same plants and have extensive interchangeability of parts and components. For example, the ZIL-131 was the main 3½-ton 6x6 Soviet military truck used in Vietnam and Afghanistan and is produced also in a civilian 4 x 2 version as the ZIL-130. Over 60 percent of the parts in the ZIL-131 military truck are common to the ZIL-130 civilian truck.

All Soviet truck technology and a large part of Soviet truck-manufacturing equipment has come from the West, mainly from the United States. While some elementary transfer-lines and individual machines for vehicle production are made in the Soviet Union, these are copies of Western machines and always obsolete in design.

Many major American companies have been prominent in building up the Soviet truck industry. The Ford Motor Company, the A. J. Brandt Company, the Austin Company, General Electric, Swindell-Dressler, and others supplied the technical assistance, design work, and equipment of the original giant plants.

This Soviet military-civilian truck industry originally comprised two main groups of plants, plus five newer giant plants. The first group used models, technical assistance, and parts and components from the Ford-built Gorki automobile plant (GAZ is the model designation). The second group of production plants used models, parts, and components from the A. J. Brandt-rebuilt ZIL plant in Moscow (Zavod imeni Likhachev, formerly the AMO and later the Stalin plant). Consequently this plant was called the BBH-ZIL plant after the three companies involved in its reconstruction and expansion in the 1930s: A. J. Brandt, Budd, and Hamilton Foundry.

---

[11]G. I. Pokrovski, *Science and Technology in Contemporary War* (New York: Frederick A. Praeger, 1959), p. 122.

There is a fundamental difference between the Ford and Brandt companies. Brandt had only one contract in the USSR, to rebuild the old AMO plant in 1929. AMO in 1930 had a production of 30,000 trucks per year, compared to the Gorki plant, designed from scratch by Ford for an output of 140,000 vehicles per year. Ford is still interested in Russian business. Brandt is not interested and has not been since 1930.

The Ford-Gorki group of assembly plants includes the plants at Ulyanovsk (model designation UAZ), Odessa (model designation OAZ), and Pavlovo (model designation PAZ). The BBH-ZIL group includes the truck plants at Mytischiy (MMZ model designation), Miass (or URAL Zis), Dnepropetrovsk (model designation DAZ), Kutaisi (KAZ model), and Lvov (LAZ model). Besides these main groups there are also five independent plants. The Minsk truck plant (MAZ) was built with German assistance. The Hercules-Yaroslavl truck plant (YaAz) was built by the Hercules Motor Company. The MZMA plant in Moscow, which manufactures small automobiles, was also built by Ford Motor Company.

In the late 1960s came the so-called Fiat-Togliatti auto plant. Three-quarters of this equipment came from the United States. Then in 1972 the U.S. government issued $1 billion in licenses to export equipment and technical assistance for the Kama truck plant. Planned as the largest truck plant in the world, it covers 36 square miles and produces more heavy trucks, including military trucks, than the output of *all* U.S. heavy truck manufacturers combined. (Togliatti and Kama are described in Chapter Three below.)

This comprises the complete Soviet vehicle manufacturing industry — **all built with Western, primarily American, technical assistance and technology.** Military models are produced in these plants utilizing the same components as the civilian models. The two main vehicle production centers, Gorki and ZIL, manufacture more than two-thirds of all Soviet civilian vehicles (excluding the new Togliatti and Kama plants) and almost all current military vehicles.

The Ford Gorki "Automobile" Plant

In May 1929 the Soviets signed an agreement with the Ford Motor Company of Detroit. The Soviets agreed to purchase $13 million worth of automobiles and parts and Ford agreed to give technical assistance until 1938 to construct an integrated automobile-manufacturing plant at Nizhni-Novgorod. Construction was completed in 1933 by the Austin Company for production of the Ford Model-A passenger car and light truck. Today this plant is known as Gorki. With its original equipment supplemented by imports and domestic copies of imported equipment, Gorki produces the GAZ range of automobiles, trucks, and military vehicles. All Soviet vehicles with the model prefix GAZ (**Gorki**

Avtomobilnyi Zavod) are from Gorki, and models with prefixes UAX, OdAZ, and PAZ are made from Gorki components.

In 1930 Gorki produced the Ford Model-A (known as GAZ-A) and the Ford light truck (called GAZ-AA). Both these Ford models were **immediately adopted for military use.** By the late 1930s production at Gorki was 80,000-90,000 "Russian Ford" vehicles per year.

The engine production facilities at Gorki were designed under a technical assistance agreement with the Brown Lipe Gear Company for gear-cutting technology and Timken-Detroit Axle Company for rear and front axles.

Furthermore, U.S. equipment has been shipped in substantial quantities to Gorki and subsidiary plants since the 1930s — indeed some shipments were made from the United States in 1968 during the Vietnamese War.

As soon as Ford's engineers left Gorki in 1930 the Soviets began production of military vehicles. The Soviet BA armored car of the 1930s was the GAZ-A (Ford Model-A) chassis, intended for passenger cars, but converted to an armored car with the addition of a DT machine gun. The BA was followed by the BA-10 — the Ford Model-A truck chassis with a mount containing either a 37-millimeter gun or a 12.7-millimeter heavy machine gun. A Red Army staff car was also based on the Ford Model-A in the pre-war period.

During World War II Gorki produced the GAZ-60 — a hybrid half-track personnel carrier that combined the GAZ-63 chassis. In the late 1940s the plant switched to production of an amphibious carrier — The GAZ-46. This was a standard GAZ-69 chassis with a U.S. quarter-ton amphibious body.

In the mid-1950s Gorki produced the GAZ-47 armored amphibious cargo carrier with space for nine men. Its engine was the GAZ-61, a 74-horsepower Ford-type 6-cylinder in-line gasoline engine — the basic Gorki engine.

In the 1960s and 1970s production continued with an improved version of the BAZ-47 armored cargo carrier, using a GAZ-53 V-8 type engine developing 115 horsepower.

In brief, the Ford-Gorki plant has a continuous history of production of armored cars and wheeled vehicles for Soviet army use: those used against the United States in Korea and Vietnam.

In addition to armored cars, the Ford-Gorki factory manufactures a range of truck-mounted weapons. This series began in the early thirties with a 76.2-millimeter field howitzer mounted on the Ford-GAZ Model-A truck. Two similar weapons from Gorki before World War II were a twin 25-millimeter antiaircraft machine gun and a quad 7.62-millimeter Maxim antiaircraft machine gun — also mounted on the Ford-GAZ truck chassis.

During World War II Gorki produced several rocket-launchers mounted on trucks. First the 12-rail, 300-millimeter launcher; then, from 1944 onwards, the M-8, M-13, and M-31 rocket-launchers mounted on GAZ-63 trucks. (The GAZ-63 is an obvious direct copy of the U.S. Army's 2½-ton truck.) Also during World War II Gorki produced the GAZ-203, 85-horsepower engine for the SU-76 self-propelled gun produced at Uralmashzavod. (Uralmash was designed and equipped by American and German companies.)

After World War II Gorki production of rocket-launchers continued with the BM-31, which had twelve 300-millimeter tubes mounted on a GAZ-63 truck chassis. In the late 1950s another model was produced with twelve 140-millimeter tubes on a GAZ-63 truck chassis. In the 1960s yet another model with eight 140-millimeter tube was produced on a GAZ-63 chassis.

Finally, in 1964 Gorki produced the first Soviet wire-guided missile antitank system. This consisted of four rocket-launchers mounted on a GAZ-69 chassis. These weapons turned up in Israel in the late 1960s. The GAZ-69 chassis produced at Gorki is also widely used in the Soviet Army as a command vehicle and scout car. Soviet airborne troops use it as a tow for the 57-millimeter antitank gun and the 14.5-millimeter double-barrelled antiaircraft gun. Other Gorki vehicles used by the Soviet military include the GAZ-69 truck, used for towing the 107-millimeter recoilless rifle (RP-107), the GAZ-46, or Soviet jeep, and the GAZ-54, a 1½-ton military cargo truck.

In brief, the Gorki plant, built by the Ford Motor Company the Austin Company and modernized by numerous other U.S. companies under the policy of "peaceful trade," is **today** a major producer of Soviet army vehicles and weapons carriers.

The A. J. Brandt-ZIL Plant

A technical assistance agreement was concluded in 1929 with the Arthur J. Brandt Company of Detroit for the reorganization and expansion of the tsarist AMO truck plant, previously equipped in 1917 with new U.S. equipment. Design work for this expansion was handled in Brandt's Detroit office and plant and American engineers were sent to Russia.

The AMO plant was again expanded in 1936 by the Budd Company and Hamilton Foundry and its name was changed to ZIS (now ZIL). During World War II the original equipment was removed to establish the URALS plant and the ZIS plant was re-established with Lend-Lease equipment.

The first armored vehicle produced at AMO was an adaptation of the civilian ZIL-6 truck produced after the Brandt reorganization in 1930. This vehicle was converted into a mount for several self-propelled weapons, including the single 76.2-millimeter antiaircraft gun and the 76.2-millimeter antitank gun.

In World War II the ZIL-6 was adapted for the 85-millimeter antitank and antiaircraft guns, quadruple 7.62 Maxims, and several self-propelled rocket-launchers, including the M-8 36-rail, 80-millimeter, and the Katyusha model M-13/A 16-rail, 130-millimeter rocket-launcher.

In the immediate postwar period the ZIL-150 truck chassis was used as a mount for the model M-13 rocket-launcher and the ZIL-151 truck was used as a mount for the M-31 rocket-launcher. In addition, the ZIL-151 truck was used as a prime mover for the 82-millimeter gun.

In 1953 the ZIL-151 truck was adapted for several other weapons, including the BM-24, 240-millimeter, 12-tube rocket-launcher; the RM-131-millimeter, 32-tube rocket-launcher; the BM-14, 140-millimeter, 16-tube rocket-launcher, and the 200-millimeter, 4-tube rocket-launcher.

In the 1960s the ZIL-157 truck became a mount for the GOA-SA-2 antiaircraft missile, and a prime mover for another rocket system.

The ZIL plant has also produced unarmored cargo and troop vehicles for the Soviet Army. In 1932 the ZIL-33 was developed; an unarmored half-track used as a troop carrier. In 1936 the ZIL-6 was developed as a half-track and during World War II the ZIL-42 was developed as a 2½-ton tracked weapons carrier. In the postwar period the ZIL-151 truck chassis was adapted for the BTR-152 armored troop carrier. In the 1950s the ZIL-485 was developed; a replica of the American DUKW mounted on a ZIL-151 truck, and followed by an improved DUKW mounted on a ZIL-157 truck.

From 1954 onwards new versions of the BTR-152 were added, based on the ZIL-157 truck. In the 1960s a new BTR-60 (8 x 8) amphibious personnel carrier was developed with a ZIL-375 gasoline engine.

Other ZIL vehicles are also used for military purposes. For example the ZIL-111 is used as a radar and computer truck for antiaircraft systems and as a tow for the M-38 short 122-millimeter howitzer The ZIL-111 is copied from Studebaker 6 x 6 trucks supplied under Lend-Lease.

There is a great deal of interchangeability between the military and civilian versions of the ZIL family of vehicles. For example, an article in *Ordnance* states:

In the 1940s the ZIL-151, a 2½-ton 6 x 6 was the work horse of the Soviet Army. It was replaced in the 1950s by the ZIL-157, an apparent product improved version. In the 1960s, however, this vehicle class requirement was met by the ZIL-131, a 3½-ton 6 x 6 vehicle, essentially a military design. It is of interest to note that a civilian version was marketed as the ZIL-130 in a 4 x 2 configuration. Over 60 percent of the components in the military version are common to the civilian vehicle.

Thus the ZIL plant, originally designed and rebuilt under the supervision of the A. J. Brandt Company of Detroit in 1930 and equipped by other American companies, was again expanded by Budd and Hamilton Foundry in 1936. Rebuilt with Lend-Lease equipment and periodically updated with late model imports, ZIL has had a long and continuous history of producing Soviet military cargo trucks and weapons carriers.

On April 19, 1972, the U.S. Navy photographed a Russian freighter bound for Haiphong with a full load of military cargo, including a deck load of ZIL-130 cargo trucks and ZIL-555 dump trucks (*Human Events, May 13, 1972*). Thus the "peaceful trade" of the 1930s, the 1940s, the 1950s, 1960s and the 1970s was used to kill Americans in Vietnam, and commit genocide in Afghanistan.

The original 1930 equipment was removed from ZIL in 1944 and used to build the Miass plant. It was replaced by Lend-Lease equipment, was supplemented by equipment imports in the 1950s, 1960s and 1970s.

The Urals plant at Miass (known as Urals ZIS or ZIL) was built in 1944 and largely tooled with equipment evacauted form the Moscow ZIL plant. The Urals Miass plant started production with the Urals-5 light truck, utilizing an engine with the specifications of the 1920 Fordson (original Ford Motor Company equipment supplied in the late 1920s was used, supplemented by Lend-Lease equipment). The Urals plant today produces weapons models: for example, a prime mover for guns, including the long-range 130-millimeter cannon, and two versions — tracked and wheeled — of a 12-ton prime mover.

Possibly there may have been doubt as to Soviet end-use of truck plants back in the 20s and 30s, but the above information certainly was known to Washington at least by the mid 1960s when this author's first volume was published. The next chapter presents official Washington's suicidal reaction to this information, under pressure from the deaf mute blindmen.

CHAPTER III

The Deaf Mutes Supply Trucks for Afghan Genocide

"The (American) businessmen who built the Soviet Kama River truck plant should be shot as traitors."

Avraham Shifrin, former Soviet Defense Ministry official

Although the military output from Gorki and ZIL was well known to U.S. intelligence and therefore to successive administrations, American aid for construction of even large military truck plants was approved in the 1960s and 1970s.

Under intense political pressure from the deaf mute blindmen, U.S. politicians, particularly in the Johnson and Nixon administrations under the prodding of Henry Kissinger (a long-time employee of the Rockefeller family), allowed the Togliatti (Volgograd) and Kama River plants to be built.

The Volgograd automobile plant, built between 1968 and 1971, has a capacity of 600,000 vehicles per year, three times more than the Ford-built Gorki plant, which up to 1968 had been the largest auto plant in the USSR.

Although Volgograd is described in Western literature as the "Togliatti plant" or the "Fiat-Soviet auto plant," and does indeed produce a version of the Fiat-124 sedan, the core of the technology is American. Three-quarters of the equipment, including the key transfer lines and automatics, came from the United States. It is truly extraordinary that a plant with known military potential could have been equipped from the United States in the middle of the Vietnamese War, a war in which the North Vietnamese received 80 percent of their supplies from the Soviet Union.

The construction contract, awarded to Fiat S.p.A., a firm closely associated with Chase Manhattan Bank, included an engineering fee of $65 million. The agreement between Fiat and the Soviet government included:

The supply of drawing and engineering data for two automobile models, substantially similar to the Fiat types of current production, but with the modifications required by the particular climatic and road conditions of the country; the supply of a complete manufacturing plant project, with the definition of the machine tools, toolings, control apparatus, etc.; the supply of the necessary know-how, personnel training, plant start-up assistance, and other similar services.

All key machine tools and transfer lines came from the United States. While the tooling and fixtures were designed by Fiat, over $50 million worth of the key special equipment came from U.S. suppliers. This included:

1. Foundry machines and heat-treating equipment, mainly flask and core molding machines to produce cast iron and aluminum parts and continuous heat-treating furnaces.

2. Transfer lines for engine parts, including four lines for pistons, lathes, and grinding machines for engine crank-shafts, and boring and honing machines for cylinder linings and shaft housings.

3. Transfer lines and machines for other components, including transfer lines for machining of differential carriers and housing, automatic lathes, machine tools for production of gears, transmission sliding sleeves, splined shafts, and hubs.

4. Machines for body parts, including body panel presses, sheet straighteners, parts for painting installations, and upholstery processing equipment.

5. Materials-handling, maintenance, and inspection equipment consisting of overhead twin-rail Webb-type conveyors, assembly and storage lines, special tool ·sharpeners for automatic machines, and inspection devices.

Some equipment was on the U.S. Export Control and Co-Com lists as strategic, but this proved no setback to the Johnson Administration: the restrictions were arbitrarily abandoned. Leading U.S. machine-tool firms participated in supplying the equipment: TRW, Inc. of Cleveland supplied steering linkages; U.S. Industries, Inc. supplied a "major portion" of the presses; Gleason Works of Rochester, New York (well known as a Gorki supplier) supplied gear-cutting and heat-treating equipment; New Britain Machine Company supplied automatic lathes. Other equipment was supplied by U.S. subsidiary companies in Europe and some came directly from European firms (for example, Hawker-Siddeley Dynamics of the United Kingdom supplied six industrial robots). In all, approximately 75 percent of the production equipment came from the United States and some 25 percent from Italy and other countries in Europe, including U.S. subsidiary companies.

In 1930, when Henry Ford undertook to build the Gorki plant, contemporary Western press releases extolled the peaceful nature of the Ford automobile, even though *Pravda* had openly stated that the Ford automobile was wanted for military purposes. Notwithstanding naive Western press releases, Gorki military vehicles were later used to help kill Americans in Korea and Vietnam.

In 1968 Dean Rusk and Walt Rostow once again extolled the peaceful nature of the automobile, specifically in reference to the Volgograd plant. Unfortunately for the credibility of Dean Rusk and Walt Rostow, there exists a proven **military** vehicle with an engine of the same capacity as the one produced at the Volgograd plant. Moreover, we have the Gorki and ZIL experience. Further, the U.S. government's own committees have stated in writing and at detailed length that **any** motor vehicle plant has war potential. Even further, both Rusk and Rostow made explicit statements to Congress denying that Volgograd had military potential.

It must be noted that these Executive Branch statements were made in the face of clear and known evidence to the contrary. In other words, the statements can only be considered as **deliberate** falsehoods to mislead Congress and the American public.

It was argued by Washington politicians that a U.S. jeep engine is more powerful than the engine built at Togliatti. The engine is indeed about two-thirds as powerful as the jeep engine, but a proven vehicle of excellent capabilities utilizing a 1,500 cc. 4-cylinder Opel engine developing 36 horsepower: this same engine later powered the Moskvitch-401 and the Moskvitch-402 (Moskva) military cross-country 4-wheel drive version of the 401, produced at the MZMA in Moscow.

In brief, there already existed a tested and usable military vehicle capable of transporting men or adaptable for weapons use and powered by a 1,500 cc. engine, the same size as the engine supplied for Togliatti. Therefore statements by U.S. officials to the effect that the Togliatti plant has no military capabilities are erroneous.

Military possibilities for such a small engine include use in a special-purpose small military vehicle (like the American jeep), or as a propulsive unit in a specially designed vehicle for carrying either personnel or weapons. Soviet strategy is currently toward supply of wars of "national liberation." The Togliatti vehicle is an excellent replacement for the bicycle used in Vietnam. The GAZ-46 is the Soviet version of the U.S. jeep, and we know that such a vehicle figures in Soviet strategic thinking.

The War Potential of the Kama Truck Plant

Up to 1968 American construction of Soviet military truck plants was presented as "peaceful trade." In the late 1960s Soviet planners decided to build the largest truck factory in the world. This plant, spread over 36 square miles situated on the Kama River, has an annual output of 100,000 multi-axle 10-ton trucks, trailers, and off-the-road vehicles. It was evident from the outset, given absence of Soviet technology in the automotive industry, that the design, engineering work, and key equipment for such a facility would have to come from the United States.

In 1972, under President Nixon and National Security Adviser Henry Kissinger, the pretense of "peaceful trade" was abandoned and the Department of Commerce admitted (*Human Events*, Dec. 1971) that the proposed Kama plant had military potential. Not only that, but according to a department spokesman, the military capability was taken into account when the export licenses were issued for Kama.

The following American firms received major contracts to supply production equipment for the gigantic Kama heavy truck plant:

Glidden Machine & Tool, Inc., North Tonawanda, New York — Milling machines and other machine tools.

Gulf and Western Industries, Inc., New York, N.Y. — A contract for $20 million of equipment.

Holcroft & Co., Kovinia, Michigan — Several contracts for heat treatment furnaces for metal parts.

Honeywell, Inc., Minneaspolis, Minnesota — Installation of automated production lines and production control equipment.

Landis Manufacturing Co., Ferndale, Michigan — Production equipment for crankshafts and other machine tools.

National Engineering Company, Chicago Illinois — Equipment for the manufacutre of castings.

Swindell-Dresser Company (a subsidy of Pullman Incorporated), Pittsburgh, Pennsylvania — Design of a foundry and equipment for the foundry, including heat treatment furnaces and smelting equipment under several contracts ($14 million).

Warner & Swazey Co., Cleveland, Ohio — Production equipment for crankshafts and other machine tools.

Combustion Engineering: molding machines ($30 million).

Ingersoll Milling Machine Company: milling machines.

E. W. Bliss Company

Who were the government officials responsible for this transfer of known military technology? The concept originally came from National Security Adviser Henry Kissinger, who reportedly sold President Nixon on the idea that giving military technology to the Soviets would temper their global territorial ambitions. How Henry arrived at this gigantic non sequitur is not known. Sufficient to state that he aroused considerable concern over his motivations. Not least that Henry had been a paid family employee of the Rockefellers since 1958 and has served as International Advisory Committee Chairman of the Chase Manhattan Bank, a Rockefeller concern.

The U.S.-Soviet trade accords including Kama and other projects were signed by George Pratt Shultz, later to become Secretary of State in the Reagan Administration and long known as a proponent of more aid and trade to the Soviets. Shultz is former President of Bechtel Corporation, a multi-national contractor and engineering firm.

American taxpayers underwrote Kama financing through the Export-

Import Bank. The head of Export-Import Bank at that time was William J. Casey, a former associate of Armand Hammer and now (1985) Director of the Central Intelligence Agency. Financing was arranged by Chase Manhattan Bank, whose then Chairman was David Rockefeller. Chase is the former employer of Paul Volcker, now Chairman of the Federal Reserve Bank. Today, William Casey denies knowledge of the military applications (see page 195), although this was emphatically pointed out to official Washington 15 years ago.

We cite these names to demonstrate the tight interlocking hold proponents of miltiary aid to the Soviet Union maintain on top policy making government positions.

On the other hand, critics of selling U.S. military technology have been ruthlessly silenced and suppressed.

Critics of Kama Silenced and Suppressed

For two decades rumors have surfaced that critics of aid to the Soviet Union have been silenced. Back in the 1930s General Electric warned its employees in the Soviet Union not to discuss their work in the USSR under penalty of dismissal.

In the 1950s and 1960s IBM fired engineers who publicly opposed sale of IBM computers to the USSR.

Let's detail two cases for the record; obviously this topic requires Congressional investigation. At some point the American public needs to know **who** has suppressed this information, and to give these persons an opportunity to defend their actions in public.

The most publicized case is that of Lawrence J. Brady, now Assistant Secretary of Commerce for Trade Administration. Ten years ago Brady was a strong critic of exporting the Kama River truck technology. In his own words (in 1982 before a Senate Investigating Committee) is Brady's view on Kama River.

Mr. Brady: Mr. Chairman, it is a privilege for me to be here again. I have testified before this subcommittee previously. As a matter of fact, it is 3 years ago this month that I testified over on the House side before the House Armed Services Committee in which I disagreed with the political appointees of the Carter administration and indicated that the technology which we were licensing to the Soviet Union, specifically for the Kama River plant, was being diverted to the Soviet military. It is 10 years ago this month that the President of the United States inaugurated the era of detente with a trip to Moscow.

A central component of that historic trip was the hope that greatly expanded trade ties between the East and the West would lead to mutual cooperation and understanding.

Obviously, those hopes have not taken place. In that 10-year period, as we in the administration have indicated in the last year, we have been exploited both legally and illegally by the Soviet Union and Eastern Europe. This technology which has helped the Soviet immensely in their military industrial infrastruture. Again, 3 years ago, I personally disclosed the failures of the Commerce Department in the licensing process, referring to it, as I said in my testimony, as a shambles.[12]

Brady went on to note that his reward for surfacing vital information was criticism and suppression.

Chairman Roth [presiding]: Thank you, Mr. Brady. Mr. Brady, the members of the subcommittee are, of course, aware of your personal commitment to this important area, but I believe it is important that the record reflect fully your position on the specific question of export technology and particularly reference the efforts some years ago to help the Soviet Union construct some trucking facilities.

Would you, for the purposes of the record, explain your role in this matter?

Mr. Brady: Mr. Chairman, about 3 years ago, the Export Administration Act was up for review for extension. As part of that review, the House Armed Services Committee decided that it was going to hold hearings on that extension, in addition to the committee of appropriate jurisdiction, namely the Foreign Affairs Committee on the House side.

There were some statements being made on both sides in Congress that were not totally consistent with the facts. We had intelligence information that trucks were being produced at the Kama River plant for the Soviet military and, in fact, being distributed to Eastern Europe for use in East European endeavors.

An administration witness was asked about that and denied it. I was asked about it and confirmed it. And, as a result of that, I was labeled a whistleblower and eventually left the Department of Commerce. In point of fact, that was the tip of the iceberg. There had been apparently intelligence through the 1970s, particularly the latter half of the seventies, indicating that there was substantial diversion taking place (and) . . . for some reason the intelligence just didn't get to the top. So that was my role. I eventually had to leave Government for it.[13]

---

[12]United States Senate, *Transfer of United States High Technology to the Soviet Union and Soviet Bloc Nations*, Hearings before the Permanent Subcommittee on Investigations, 97th Congress Second Session, May 1982, Washington, D.C., p. 263.

[13]Ibid., pp. 267-8.

However, Mr. Brady was unaware of a similar and much earlier story of suppression in the Kama case which paralleled his own.

In the years 1960-1974 this writer authored a three volume series, *Western Technology and Soviet Economic Development*, published between 1968 and 1973 by the Hoover Institution, Stanford University, where the author was Research Fellow. This series cataloged the origins of Soviet technology from 1917 down to the early 1970s. The series excluded the military aspects of technical transfers. However, the work **totally** contradicted U.S. Government public statements. For example, in 1963 State Department claimed in its public pronouncements that **all** Soviet technology was indigenous, a clear misunderstanding or dismissal of the facts.

By the early 1970s it was clear to this author that a significant part of Soviet military capability also came from the West, even though this assessment was also refuted by U.S. government analysts. Quietly, without government or private funding, this author researched and wrote *National Suicide: Military Aid to the Soviet Union*. The manuscript was accepted by Arlington House. Both author and publisher maintained absolute silence about the existence of the manuscript until publication date.

When news of publication reached Stanford, there was immediate reaction — a hostile reaction. A series of meetings was called by Hoover Institution Director W. Glenn Campbell. Campbell's objectives were:

1) to withdraw the book from publication,
2) failing that, to disassociate Hoover Institution from the book and the author.

Campbell initially claimed that *National Suicide* was a plagiarism of the author's works published by Hoover. This was shown to be nonsense. In any event an author can hardly plagiarize himself. The objective, of course, was to persuade author and publisher to withhold publication. Both the author and Arlington House refused to withdraw the book and continued with publication. The book was published and sold over 50,000 copies.

After the unsuccessful attempt at suppression Glenn Campbell arbitrarily removed the title Research Fellow from the author and removed both his name and that of his secretary from the personnel roll of the Hoover Institution. This effectively disassociated Hoover Institution from the book and its contents. The author became a non-person. Two years later the author voluntarily left Hoover Institution and assumed a private role unconnected with any research foundation or organization. These events happened some years before Mr. Brady of Commerce took his own personal stand and suffered a similar fate.

By a strange quirk of fate, Glenn Campbell is today Chairman of Mr. Reagan's Intelligence Oversight Committee.

Who were the Deaf Mute Blindmen at Kama River?

Clearly, the Nixon Administration at the highest levels produced more than a normal number of deaf mutes — those officials who knew the story of our assistance to the Soviets but for their own reasons were willing to push forward a policy that could only work to the long run advantage of the United States. It is paradoxical that an Administration that was noisy in its public anti-communist stance, and quick to point out the human cost of the Soviet system, was also an Administration that gave a gigantic boost to Soviet military truck capacity.

Possibly campaign contributions had something to do with it. Multinationals listed below as prime contractors on Kama River were also major political contributors. However, the significant link never explored by Congress is that Henry Kissinger, the key promoter of the Kama River truck plant at the policy level, was a former and long-time employee of the Rockefeller family — and the Rockefellers are the largest single shareholders in Chase Manhattan Bank (David was then Chairman of the Board) and **Chase was the lead financier for Kama River.** This is more than the much criticised "revolving door." It is close to an arm's length relationship, i.e., the use of public policy for private ends.

Here are the corporations with major contracts at Kama River, listed with the name and address of the Chairman of the Board in **1972.**

GULF & WESTERN INDUSTRIES, INC.
1 Gulf and Western Plaza, New York NY 10023
Tel. (212) 333-7000
Chairman of the Board: Charles G. Bluhdorn
Note: Charles Bluhdorn is also a Trustee of Freedoms Foundations at Valley Forge and Chairman of Paramount Pictures Corp.

E. W. BLISS CO. (a subsidiary of Gulf & Western)
217 Second Street NW, Canton, Ohio 44702
Tel. (216) 453-7701
Chairman of the Board: Carl E. Anderson
Note: Carl E. Anderson is also Chairman of the American-Israel Chamber of Commerce & Industry

COMBUSTION ENGINEERING, INC.
277 Park Avenue, New York, NY 10017
Tel. (212) 826-7100
Chairman of the Board: Arthur J. Santry, Jr.

HOLCROFT AND COMPANY
12062 Market Street, Livonia, Mich. 48150
Tel. (313) 261-8410
Chairman of the Board: John A. McMann

HONEYWELL, INC.
2701 4th Avenue S., Minneapolis, Minn. 55408
Tel. (612) 332-5200
Chairman of the Board: James H. Binger
INGERSOLL MILLING MACHINE COMPANY
707 Fulton Street, Rockford, ILL 61101
Tel. (815) 963-6461
Chairman of the Board: Robert M. Gaylord
NATIONAL ENGINEERING CO.
20 N. Wacker Drive, Chicago, ILL 60606
Tel. (312) 782-6140
Chairman of the Board: Bruce L. Simpson
PULLMAN, INC.
200 S. Michigan Ave., Chicago, ILL 60604
Tel. (312) 939-4262
Chairman of the Board: W. Irving Osborne, Jr.
SWINDELL-DRESSLER CO. (Division of Pullman, Inc.)
441 Smithfield Street, Pittsburgh, PA 15222
Tel. (412) 391-4800
Chairman of the Board: Donald J. Morfee
WARNER & SWAZEY
11000 Cedar Avenue, Cleveland, Ohio 44106
Tel. (216) 431-6014
Chairman of the Board: James C. Hodge
CHASE MANHATTAN BANK
Chairman of the Board: David Rockefeller

CHAPTER IV

Soviets Buy into the 21st Century

In most fields of technical research, development and production which I am familiar with in the Soviet Union, the overwhelming majority of resources are invested in military applications . . . as a matter of fact the Soviet industrial capacity is so overburdened with military production that the Soviets could not make a civilian or commercial application of certain high technology products even if they wanted to.

> Former Soviet engineer, Joseph Arkov
> before U.S. Senate, May 4, 1982.

Every generation or so in the past two hundred years Western technology has generated a fundamental innovation that changes the whole course of society and the economy. The industrial revolution of the late 18th and 19th centuries was based on canals and iron. Railroads were a fundamental innovation of the first third of the 19th century. In the late 19th century the Bessemer process enabled mass production of cheap steel. The internal combustion engine in the 1900s began another revolution. Atomic energy in the 1940s started the atomic age.

In the 1970s the semi-conductor was first mass produced in California. The economy of the 21st century will evolve around the silicon chip, i.e., the integrated circuit memory chip and semi-conductor components.

No country large or small will make any progress in the late 20th century without an ability to manufacture integrated circuits and associated devices. These are the core of the new industrial revolution, both civilian and military, and essentially the same device is used for both military and civilian end uses. A silicon chip is a silicon chip, except that military quality requirements may be more strict than civilian ones.

This electronic revolution originated in Santa Clara Valley, California in the 1950s and roughly centers around Stanford University.

Stanford is also in many ways at the core of the debate over transfer of our military technology to the Soviet Union. Congressman Ed Zschau (Rep. Menlo Park) represents the Silicon Valley area and is a strong proponent of more aid to the Soviets. On the other hand, also in Silicon Valley, this author's six books critical of our technological transfers to the Soviets originated, and three were published at the Hoover Institution, Stanford University (See Bibliography for titles).

Silicon Valley gets its name from the essential element silicon used in integrated circuits. An essential component of integrated circuits is the semi-conductor usually made of silicon and linked to other components such as transistors into a single circuit. By 1971 an entire computer could be produced on a single chip, in itself probably the most significant industrial breakthrough since the discovery that steel could be manufactured on a large scale from iron.

The semi-conductor revolution began in the Silicon Valley and was a challenge to the socialist world to duplicate. This they could not do. Every single Soviet weapon system has semi-conductor technology which originated in California and which has been bought, stolen or acquired from the United States.

Early Soviet Electronic Acquisitions

Back in 1929 *Pravda* commented that without the automobile the Soviet Army would be helpless in any future war. Western multinationals Ford Motor Company, Hercules Gear, IBM and others helped USSR bridge the gap of the 1920s. Identical aid can be found for electronics.

In August 1971 the U.S. Department of Defense paid $2 million to Hamilton Watch Company for precision watchmaking equipment. Watchmaking equipment is used in fabricating bomb and artillery shell fuses, aircraft timing gear, pinions, and similar military components. Most Soviet watch-manufacturing equipment has been supplied from the United States and Switzerland; in some cases the Soviets use copies of these foreign machines.

In 1929 the old Miemza concession factory, formerly a tsarist plant, received the complete equipment of the Ansonia Clock Company of New York, purchased for $500,000. This became the Second State Watch Factory in Moscow, brought into production by American and German engineers, and adapted to military products.

In 1920 the complete Deuber-Hampton Company plant at Canton, Ohio, was transferred to the Soviet Union, and brought into production by forty American technicians. Up to 1930 all watch components used in the Soviet Union had been imported from the United States and Switzerland. This new U.S.-origin manufacturing capability made possible the production of fuses and precision gears for military purposes; during World War II it was supplemented by Lend-Lease supplies and machinery.

After World War II Soviet advances in military instrumentation were based on U.S. and British devices, although the German contribution was heavy in the 1950s. About 65 percent of the production facilities removed from Germany were for the manufacture of power and lighting equipment, telephone, telegraph, and communications equipment, and cable and wire. The remainder consisted of German plants to manufacture radio tubes and radios, and military electronics facilities for such items as secret teleprinters and anti-aircraft equipment.

Many German wartime military electronic developments were made at the Reichpost Forschungsinstitut (whose director later went to the USSR) and these developments were absorbed by the Soviets, including television, infrared devices, radar, electrical coatings, acoustical fuses, and similar equipment. But although 80 percent of the German electrical and military electronics industries were removed, the Soviets did not acquire modern computer, control instrumentation, or electronic technologies from Germany: these they acquired from the U.S.

Bridging the Semi-conductor Gap

Taking semi-conductors as an example, three stages can be identified in the transfer process. The Soviets were able to import or manufacture small laboratory quantities of semi-conductors from an early date. What they could **not** do, as in many other technologies, was mass produce components with high quality characteristics. This situation is described by Dr. Lara Baker, a Soviet computer expert, before Congress:

> The Soviet system in preproduction can manage to produce a few of almost any product they want, provided they are willing to devote the resources to it. The best example of this would be the Soviet 'civilian' space program, in which they managed to put people in orbit before the United States did, but at a high cost.
>
> In the area of serial production, that is, the day to day production of large quantities of a product, the differences between the two systems become most obvious . . . Serial production is the Achilles heel of the Soviet bloc. Especially in high technology areas, the big problem the Soviets have is quality assurance . . . they count products, not quality products. This is the area where the Soviets exhibit weakness and need the most help.[14]

The first phase for the Soviets was to **identify** the technology needed, in this case a semi-conductor plant, to bridge the chasm between the 19th century and the 21st century.

The second phase was to obtain the **equipment** to establish a manufacturing plant.

The third phase was to bring this plant into production and make the best use of its output in an economy where developmental engineering resources do not exist in depth and military objectives have absolute priority.

We shall demonstrate in Chapter Five how the Soviets achieved the first of these tasks — with the help of the Control Data Corporation, Mr. William Norris, Chairman. The second phase was achieved through an illegal espionage network, the Bruchhausen network. The third phase is

[14]United States Senate, *Transfer of United States High Technology to the Soviet Union and Soviet Bloc Nations* Hearings before the Permanent Subcommittee on Investigations, 97th Congress Second Session, May 1982, Washington, D.C., p.

today in progress, although the phases one and two are already in place in the Soviet military complex.

The emphasis in this critical transfer of semi-conductor technology was not reverse engineering as, for example, the Soviet Agatha computer is reverse engineered from the Apple II computer, but use of U.S. manufacturing techniques and equipment to bridge a gigantic gap in Soviet engineering capabilities. The Soviet system does not generate the wealth of technology common in the West. It cannot choose the most efficient among numerous methods of achieving a technical objective because the Marxist system lacks the abundant fruits of an enterprise system. The emphasis in semi-conductors is transfer of a complete manufacturing technology to produce high quality products for known military end uses WHICH COULD NOT HAVE BEEN ACHIEVED BY THE SOVIETS THEMSELVES, WITHOUT FUNDAMENTAL CHANGES IN THEIR SYSTEM.

In brief, in electronics the key is not copying Western technology as for example the Caterpillar tractor was duplicated by the millions, but to transfer specialized production equipment to mass produce critical components.

This assertion has been fully supported by expert witnesses before Congressional committees. For example, the following statement was made in 1982 to the Senate by Dr. Stephen D. Bryen, Deputy Assistant Secretary of Defense, International Economics, Trade and Security Policy.

> **Senator Nunn:** Dr. Bryen, you made reference in your testimony to the effort by the Soviets to build and equip a semiconductor plant using equivalent know-how from the United States.
>
> Could the Soviets have built and equipped such a plant in the late 1970s and early 1980s without U.S. machinery, equipment and know-how?
>
> **Dr. Bryen:** My answer is they could not. That doesn't mean that equipment necessarily came from this country. It could have been transferred from Europe or elsewhere. In fact, it could have been transferred from another country that bought that equipment — it could have been on the secondary market. There is a secondary market in this sort of machinery. These are terribly difficult things to trace.
>
> What we know in the first instance is that a lot had to be U.S. equipment, that the system was full of holes, it was porous, it was easy for them to get it and they got it.
>
> The microelectronics area has enabled the Soviets to upgrade their military equipment.[15]

And a similar comment from a former Russian engineer, Joseph Arkov, again in Senate testimony.

By using — not copying — the American high technology products, they move closer to their goal of technical self-sufficiency. Whether they will ever become self-sufficient in high technology is a debatable point. My own view is that this course of action gives them quick gains, but over the long run, it will result in their being permanently behind the United States, forever having to rely on American products to manufacture their own.

However, being behind us in technology is a relative condition. The Soviets can make progress in a technical sense and, at the same time, trail the United States, but by their standards, they will have achieved much. Their accomplishments will have been made with limited cost to them because the basic research and development will have been paid for by the Americans.

To repeat, then, the Soviet strategy in obtaining American high technology products includes efforts to copy and duplicate, but the Soviets' primary objective is to obtain machinery which they can use in the manufacture of their own high technology equipment.

This distinction — the difference between copying of technology and the use of it — is an important one because it provides the United States with a key insight into which products the Soviets are the most anxious to obtain. It also can influence American policymakers in deciding which products the United States can afford to sell the Soviet Union, and which components should not be sold to them.

Soviet strategy in using American products can be seen in the following illustration. Let us say, for example, that the Soviets have 100 plants involved in producing components for use in space flight. Each of the plants could use a certain kind of American computer. But they cannot obtain 100 computers; that is, one for each plant. Instead, they are able to obtain three or four American computers of the desired type. They use the computers as best they can in those three or four plants where they can do the most good. They are not inclined to use them as non-producing models to be studied in a laboratory for the purpose of copying.

Moreover, if the American product obtained in another transaction — if, for example, the product is a sophisticated oven used in the heating of microchips — then they are even less interested in

[15]United States Senate, *Transfer of United States High Technology to the Soviet Union and Soviet Bloc Nations* Hearings before the Permanent Subcommittee on Investigations, 97th Congress Second Session, May 1982, Washington, D.C., p. 259.

copying or imitating. They will use the oven to produce microchips. There is no civilian use for equipment used to manufacture integrated circuits or semi-conductors.[16]

How the Deaf Mute Blindmen Helped the Soviets into the 21st Century

With these insights into Soviet technological acquisition strategy we can identify the stages by which the Soviets acquired semi-conductor technology. Chronologically these are:

DATE	EVENT
1951	Semi-conductor developed in Santa Clara Valley, California. From this point on Soviets import chips and then manufacture on a laboratory scale.
1971	"Computer in a chip" development. Soviets still unable to mass produce even primitive semi-conductor devices.
1973	Control Data Corporation (CDC) agrees to supply Soviets with a wide range of scientific and engineering information including construction and design of a large fast computer (75 to 100 million instructions per second is fast even in 1985) and manufacturing techniques for semi-conductors and associated technologies (See Chapter Five).
1977-80	Soviets acquire technology for a semi-conductor plant through the Bruchhausen network and Continental Trading Corp. (CTC). The CDC agreement gives Soviets sufficient information to set up a purchasing and espionage program. CDC told the Soviets what they needed to buy.
1981-82	Commerce Department lax in enforcing export control regulations. U.S. Customs Service makes determined efforts to stop export of semi-conductor manufacturing equipment.
1985	Soviets establish plant for semi-conductor mass production. Soviet military equipment based on this new output.
1986	U.S. taxpayer continues to support a defense budget of over $300 billion a year. Without these transfers Soviet military could not have been computerised and U.S. defense budget reduced.

The Bruchhausen Network

The second phase of the acquisition of semi-conductor mass production technology was the Bruchhausen network.

This network comprised a syndicate of 20 or so "front" electronics companies established by Werner J. Bruchhausen, age 34, a West Ger-

[16]United States Senate, *Transfer of United States High Technology to the Soviet Union and Soviet Bloc Nations* Hearings before the Permanent Subcommittee on Investigations, 97th Congress Second Session, May 1982, Washington, D.C., p. 29.

man national. The key component was a group of companies with the initials CTC (Continental Trading Corporation), managed by Anatoli Maluta, a Russian-born naturalized U.S. citizen. A Congressional sub-committee devoted considerable time and resources to reconstruction of the activities of the CTC-Maluta operation.

This network of companies, controlled from West Germany, gave the Soviets the technology for a major leap forward in modernizing military electronics capability. Dr. Lara H. Baker, Jr., who had personal knowledge of the CTC-Maluta case, was one of the subcommittee's sources in reconstructing the network. Other sources included the Departments of Commerce and Justice and the U.S. Customs Service.

Using Werner Bruchhausen's companies and accomplices in Western Europe as freight forwarders and transshipment points, Maluta sent more than $10 million of American-made high technology equipment to the Soviet Union from 1977 to 1980. the machinery was used to equip a Soviet plant for the manufacture and testing of semi-conductors. The equipment went from California to Western Europe to the USSR.

To Dr. Baker, the CTC-Maluta case proved a point: that the Soviets know precisely what U.S. technology they want, and leave little to chance. Dr. Baker explained:

Of particular interest to me in the (CTC-Maluta) case is the in-formation it gives us about Soviet intentions. We delude ourselves if we think the Soviets enter the black market in search of strategic components in a helter-skelter style, buying up dual-use com-modities without rhyme or reason.

The truth of the matter is that the Soviets and their surrogates buy nothing they don't have specific, well defined needs for. They know exactly what they want — right down to the model number — and what they want is part of a carefully crafted design.

The carefully crafted design in this instance was a semi-conductor manufacturing plant, an essential part of the Soviet need to close the technological gap between themselves and the U.S. in the integrated circuit/microcomputer industry. We shall see in Chapter Five how Con-trol Data Corporation provided the key basic information on **what** to acquire.

Dr. Baker, who testified in the 1981 successful prosecution of Maluta and his associate, Sabina Dorn Tittel, studied 400 separate air waybills and other shipping documents used by the CTC network. The conclu-sion was inescapable that the Soviets were equipping a semi-conductor plant. Soviet use of components of U.S. origin demonstrated their determination to make the facility as efficient and modern as any in the world:

. . . (the Soviets) have purchased clandestinely all the hardware they need for equipping a good integrated circuit production plant. They showed no interest in purchasing production equipment that was not state of the art. They showed very good taste.

Stressing the point that, through the CTC-Maluta combine, the Soviets bought everything needed for a semi-conductor manfacturing plant, Dr. Baker testified to the Senate that among the equipment bought in the period 1977 through 1980 were saws for cutting silicon crystals, equipment for making masks for integrated circuit production, plotters to draw the circuits, basic computer-aided design systems for integrated circuit design, diffusion ovens for circuit production ion-implantation systems for circuit production, photo-lithographic systems for integrated circuit production, scribers for separating integrated circuits on wafers, testers for testing integrated circuits on wafers, bonding equipment for bonding connecting leads to integrated circuits, and packaging equipment for packaging the circuits.

Dr. Baker added:

High quality integrated circuits are the basis of modern military electronics. Integrated circuits form the basis for military systems which are more flexible, more capable and more reliable than systems using discrete electronic components. The production tooling and equipment obtained by the Soviets (from the CTC-Maluta network) will significantly improve the Soviets' capability to produce such circuits.

Further support for the assertion that the Soviets relied on American technology to equip their semi-conductor plant came from John D. Marshall, a chemist and specialist in facilities that manufacture semiconductors.

Marshall owns a high technology business in Silicon Valley and testified to Congress that in the winter of 1975 he made two trips to the Soviet Union. Led by a West German named Richard Mueller to believe that the Soviets wanted to retain his consultative services in connection with their plans to manufacture electronic watches, Marshall learned on the second trip to Moscow that what was actually wanted was expertise to equip a semi-conductor plant. Marshall told the subcommittee:

On the second trip, we met several Soviets who purported to be technical people. They were not very well trained and were not familiar with sophisticated technological thinking. But it was apparent to me by the questions they asked and the subjects they discussed that the Soviets had built a semi-conductor manfuacturing and assembly plant and they were anxious to equip it.

They wanted American semi-conductor manufacturing equipment and they had detailed literature on the precise kind of equipment they wanted. They also wanted me to obtain for them certain semi-conductor components.

It was clear to me that Mueller had deceived me as to the Soviets' intentions, that it was not merely electronic watches the Soviets wanted to manufacture.

Marshall realized that to cooperate further with the Soviets would be illegal. He refused to meet further with the Soviets and left Moscow.

As he returned to the United States, Marshall recalled conversations he had overheard that at the time had not made sense to him on the way to Moscow. Marshall and Mueller had stopped in Hamburg where Mueller introduced him to a Canadian, also providing technical assistance to the Soviets, that his mission was to show them how to make integrated circuits and to use equipment now on the way.

In Moscow, Marshall said, he met a woman who spoke English with a German accent who planned to ship certain American-made photo-lithography materials to the Soviet Union via East Berlin. Photo-lithography materials are critical in semi-conductor manufacture.

In West Germany, Marshall was introduced to Volker Nast, identified by Mueller as his partner. Nast was involved in illegal diversions of the U.S. technology to the Soviet Union.

The significance of 1975 as the year the Soviets expressed their desire for American-made semi-conductor equipment has been explained by Marshall. In 1975 the U.S. was preeminent in the field of semi-conductor technology. Marshall said:

> It is my view that the Soviets had built their manufacturing plant, or plants, to specifications for American-made equipment — for the manufacture, assembly and testing of integrated circuits. Now that the facilities were constructed, they were, in the winter of 1975, confronted with the next step, which was to equip the facilities.

According to Marshall, the Soviets' primary interest in 1975 was the manufacture and assembly phases of semi-conductor production. By 1977, he said, the Soviets needed to stock the facility with test equipment, and software development equipment.

Dr. Lara Baker, in his testimony before Congress, agreed with Marshall. In the 1978-79 time frame the CTC-Maluta syndicate purchased production equipment. In the 1979-1980 period, the CTC-Maluta network bought semi-conductor test equipment. Said Baker, "Marshall's testimony is quite consistent with my information."

U.S. Customs Service investigations confirmed not only the Bruchhausen network but subsidiary networks operating in cooperation with the Soviets for illegal purchase of semi-conductor manufacturing equipment.

Information about the Soviets' efforts to build a semi-conductor industry — and, in so doing, make a major leap forward in military electronics — was given to a Senate subcommittee by Charles L. McLeod, a Special Agent with the U.S. Customs Service. McLeod said the same Richar Mueller who had brought John Marshall to Moscow had been active in several other schemes.

In fact, McLeod said, Mueller was an operative in a syndicate whose mission was to export by illegal means semi-conductor manufacturing equipment from the United States to the Soviet Union. Other operatives in the network included Volker Nast, Luther Heidecke, Peter Gessner and Frederick Linnhoff, all West Germans. In the U.S., their associates included Robert C. Johnson, Gerald R. Starek and Carl E. Storey, officers in high technology firms.

McLeod, of the San Francisco office of U.S. Customs, investigating technology diversions originating in nearby Santa Clara County, said inquiries into two Silicon Valley electronics firms — II Industries and Kaspar Electronics — led to the conclusion that the Soviets were trying in the mid-1970s to "construct a semi-conductor manufacturing facility by using U.S. technology and equipment." A loosely knit organization of electronics producers and brokers in West Germany and Northern California assisted the Soviets.

The first diversion of semi-conductor manufacturing equipment occurred in 1974. McLeod said participants in the diversion included Luther Heidecke, a representative of Honeywell/West Germany, AG, and Peter Gessner, the European sales representative for Applied Materials, a Northern California firm which produced semi-conductor manufacturing equipment. Gessner had other jobs as well, serving as the European salesman for II Industries and Kaspar Electronics. In addition, Gessner was employed by Richard Mueller.

Processing orders through Honeywell/West Germany for the purchase of semiconductor manufacturing equipment, Heidecke arranged for the export of II Industries and Kaspar Electronics machinery to West Germany and ultimately to the Soviet Union. Heidecke's activities came to the attention of the West German authorities, who prosecuted him for giving the Soviets national security information.

McLeod described a second diversion. A Mays Landing, New Jersey export firm known as Semi-Con, formed by a West German Richard Mueller and managed by a former intelligence agent, shipped semiconductor manufacturing equipment to the Soviet Union. The equipment was from II Industries and Kaspar Electronics.

The identification of the Bruchhausen network does not reflect favorably upon the operating efficiency of the Commerce Department's Office of Export Control and specifically its Compliance Division. (As the situation existed in 1980, it may possibly have improved since that time.)

The existence of the CTC network of companies was first brought to the attention of the U.S. government in 1977 and 1978 when two anonymous letters were received at the American Consulate in Dusseldorf, Germany. The State Department translated the letters into English and referred them to the Compliance Division in Commerce. The letters were received by the Compliance Division in 1978 and no effort was made to investigate the allegations.

After receipt of the letters, two U.S. producers of dual-use technology also reported to the Commerce Department that they were suspicious of the CTC companies. Again no Commerce action.

A Commerce Department special agent did interview CTC's principal executive in Los Angeles, the naturalized Russian-born American citizen Anatoli Maluta, also known as Tony Maluta and Tony Metz. Maluta told the special agent from Compliance that he did not know anything about export controls, or the need to have validated export licenses to ship certain controlled commodities. However, Maluta said, because of the agent's interest, he would cancel the suspicious order.

There was no further investigation of the CTC network until a second letter arrived at Compliance Division from another high-technology producer, also suspicious about the CTC companies.

Early in 1980, a second Compliance Division agent, Robert Rice, was assigned to the case and conducted a comprehensive preliminary inquiry. Rice, the most senior agent in the Division, found considerable information indicating widespread violations of export controls. The evidence went to the Office of the U.S. Attorney in Los Angeles in March 1980. A major inquiry was begun by the U.S. Attorney's office, under the direction of Assistant U.S. Attorney Theodore W. Wu and the U.S. Customs Services. Customs ultimately assigned about 15 agents to the case in California, Texas, New York, and Western Europe. Compliance Division Special Agent Rice was the only Commerce Department representative assigned to the case on a regular basis.

Indictments were brought against Bruchhausen and Dietmar Ulrichshofer, both of whom remained in Europe out of reach, and two Los Angeles associates — Maluta and Sabina Dorn Tittel. Maluta and Tittel were convicted.

The CTC case demonstrated technology diversions of about $10 million and is considered by law enforcement and national security specialists one of the most important export control cases ever brought to trial. The inquiry showed that:

First, the Compliance Division did not move quickly to establish the value of the anonymous letters.

Second, the Compliance Division did not connect the anonymous letters to the allegations reported by two U.S. manufacturers.

Third, when Compliance Division Agent Rice turned over the results of his inquiry to Assistant U.S. Attorney Wu in Los Angeles, it was apparent to Wu that considerable expenditures of resources would be needed. Trained investigators would be required to conduct interviews, evaluate shipping documents, surveil suspected violators, and carry out other aspects of a traditional law enforcement full-scale field investigation.

Commerce's contribution to that effort was Agent Rice, a competent investigator in whom Wu had confidence. But he needed more than one agent, and enlisted assistance of the Customs Service. Later assistance was provided by trained criminal investigators from the IRS.

Fourth, at an early point in the inquiry it was necessary to seize shipments. Commerce had neither the authority nor the manpower to seize shipments. Customs did.

Fifth, at another point in the inquiry it was necessary to search the premises of CTC companies and certain of employees in the United States and Europe. The Compliance Division had insufficient resources to implement simultaneous search warrants. The Compliance Division had no law enforcement capabilities in Western Europe to work with German customs to coordinate searches abroad. Customs executed the warrants in the United States and, through its agreements with West German customs, arranged for the execution of warrants in Germany.

Sixth, to substitute sand for one of CTC's shipments to Moscow, a sizable expenditure of funds was needed. The Compliance Division balked at the shipment substitution strategy and refused to pay the cost of recrating the sand and airfreight. Customs officials approved of the substitution and agreed to pay the cost.

Seventh, extensive overseas coordination, in addition to the search warrants, was called for with West German Customs and other overseas law enforcement contacts.

Eighth, extensive surveillance was necessary. Armed Customs agents and armed Internal Revenue Service criminal investigators and an unarmed Compliance Division Special Agent Rice provided this work.

Two suspects under surveillance had firearms in the back seat of their car. The firearms were not used. But it was an important law enforcement advantage for the agents on surveillance to be armed as well.

Ninth, experienced supervisors with law enforcement background and training were needed to direct the inquiry in the field. The Office of the U.S. Attorney for the Central District of California, working with supervisorial personnel in the Customs Service, provided the needed direction. Contact with supervisorial personnel in the Compliance Division, who remained in Washington, was made on the telephone and the persons who worked on the case in California did not consider such communication to be satisfactory.

Tenth, when the appropriate time came to apprehend Anatoli Maluta and Sabina Dorn Tittle, IRS agents made the arrests. Customs agents, like the IRS criminal investigators, are authorized to make arrests. Even had the Compliance Division dispatched sufficient numbers of agents to assist in the inquiry, they could not have arrested the suspects.

The Type of Equipment Shipped to the USSR

The Bruchhausen network was extremely efficient at obtaining the type of equipment needed by the Soviets for their semiconductor plant.

Here is a summary as reported to a Congressional Committee by an American computer expert.

Senator Nunn: Were you familiar with the nature of the equipment that was being shipped for the Soviets?

Mr. Marshall: Yes, I was.

Senator Nunn: What was it to have been used for in your opinion?

Mr. Marshall: It would have been used for the production of the integrated circuits. It was part of the photolithography process used to make integrated circuits.

Senator Nunn: What about the military applications?

Mr. Marshall: The circuits certainly have military application; the equipment has no military application.

Senator Nunn: This is primarily industrial?

Mr. Marshall: For production circuits used to print the patterns, the microscopic patterns on the silicon —

Senator Nunn: Does that mean once they produce these they would have been using it for commercial purposes?

Mr. Marshall: Commercial or military.

Senator Nunn: Or military?

Mr. Marshall: Yes.

Senator Nunn: Members of the minority staff showed you a list of equipment that has been illegally exported to the Soviets over the period 1976 to 1980; is that correct?

Mr. Marshall: Yes.

Senator Nunn: These illegal exports were valued at about $10 million. Have you looked at that list?

Mr. Marshall: I have seen the list, yes.

Senator Nunn: What type of equipment was this and how would it have been used, in your opinion?

Mr. Marshall: Most of the equipment there really broke down into two categories. One category was mainly test equipment, for testing integrated circuits. Another area was software development.

Examination of Continental Technology Corporation invoices demonstrates the high technology specialized nature of the equipment shipped by the Bruchhausen network. Here are sample summaries based on CTC invoices:

CTC Invoice Number:　1017
Manufacturer:　　　　Fairchild Instrument Corporation

The commodities on this invoice are spare parts and extensions for the Fairchild Xincom test systems referred to on CTC Invoice number 8051, among others. As such they continue and enhance the capabilities of the microprocessor test system previously referred to.

The products also exceed the state of the art of new designs in the destination country, as of the time of shipment.

CTC Invoice Number:　8071
Manufacturer:　　　　Fairchild Instrument Corporation

The commodity on this invoice is a test system for the production testing of integrated circuits. This test system is necessary and critical for the test of civilian or military microcircuits. This is an area in which the destination country falls far behind the United States in capability, and the test system is applicable to integrated circuits for military applications. Single board computers/microcomputer systems are used in most advanced weapons systems throughout the free world, particularly in missile and aircraft systems. Their use provides a significant increase in effectiveness with an equally significant reduction in weight and power consumption.

These products would exceed the state of the art of equipment being manufactured in the destination country, as of the time of shipment. The products also exceed the state of the art of new designs in the destination country, as of the time of shipment.

CTC Invoice Number:　1091
Manufacturer:　　　　Fairchild Instrument Company

The commodity on this invoice is a test system for the production testing of integrated circuits. This test system is necessary and critical for the test of civilian or military microcircuits. This is an area in which the destination country falls far behind the United States in capability and the test system is applicable to integrated circuits for military applications. Single board computers/microcomputer systems are used in most advanced weapons systems throughout the free world, particularly in missile and aircraft systems. Their use provides a significant increase in effectiveness with an equally significant reduction in weight and power consumption.

These products would exceed the state of the art of equipment being manufactured in the destination country, as of the time of shipment. The products also exceed the state of the art of new designs in the destination country, as of the time of shipment.

CTC Invoice Number: 1040
Manufacturer: Fairchild Instrument Corporation

The commodity on this invoice is a test system for the production testing of integrated circuits. This test system is necessary and critical for the test of civilian or military microcircuits. This is an area in which the destination country falls far behind the United States in capability, and the test system is applicable to integrated circuits for military applications. Single board computers/microcomputer systems are used in most advanced weapons systems throughout the free world, particularly in missile and aircraft systems. Their use provides a significant increase in effectiveness with an equally significant reduction in weight and power consumption.

These products would exceed the state of the art of equipment being manufactured in the destination country, as of the time of shipment. The products also exceed the state of the art of new designs in the destination country, as of the time of shipment.

CTC Invoice Number: 21 073
Manufacturer: California Computer Products Incorporated

The commodity on this invoice is a complete off-line high-precision flat bed plotter system. The plotter involved, the CALCOMP-748 plotter, is precise enough and big enough to directly draw the masks needed for integrated circuit production. For that reason the plotter associated equipment is embargoed. The integrated circuits produced with masks drawn on this plotter can be used for civilian or military applications and as such are embargoed.

The products exceed the state of the art of new designs in the destination country, as of the time of shipment.

CTC Invoice Number: 21 074
Manufacturer: California Computer Products Incorporated

The commodities on this invoice are accessories for the high precision plotter referred to on CTC Invoice number 21 073. As such they are embargoed for that application. These particular pieces of equipment are within the state of the art of the country of destination, however, they could not be shipped as part of the plotter system.

CTC Invoice Number: 21 075
Manufacturer: California Computer Products Incorporated

The commodities on this invoice are spare parts for the plotter referred to on CTC Invoice number 21 073. As such they would be embargoed because of the direct military application of the plotter. The parts themselves may be embargoed because they contain embargoed technology. These products would exceed the state of the art of equipment being manufactured in the destination country as of the time of shipment.

CTC Invoice Number: 22 004
Manufacturer: Tektronics Incorporated

The commodity on this invoice is an extremely high speed (350 megahertz) oscilloscope with direct military applications in nuclear weapons testing, in high speed signal processing systems, and in other high speed electronic applications. This product exceeds the state of the art of equipment being manufactured in the destination country as of the time of the shipment.

The product matches the state of the art of new designs in the destination country, as of the time of shipment.

CTC Invoice Number: 1188
Manufacturer: Tektronics Incorporated

The commodities on this invoice are accessories for the high speed oscilloscope referred to on CTC Invoice 22 004. As such they are embargoed. These products exceed the state of the art of equipment being manufactured in the destination country as of the time of shipment. The products also exceed the state of the art of new designs in the destination country, as of the time of shipment.

CTC Invoice Number: 1003
Manufacturer: Data General Corporation

The commodity on this invoice is a complete Data General Eclipse S/230 digital computer with substantial peripheral and input/output communication equipment. This general purpose computer could be licensed if a license were applied for and certain characteristics of the computer were deleted. As a general purpose computer it is applicable to many civilian and military applications. This particular configuration seems applicable to the control and monitoring of the manufacture of integrated circuits.

The products also exceed the state of the art of new designs in the destination country, as of the time of shipment.

CTC Invoice Number: 21089
Manufacturer: Data General Corporation

So confident are the Soviets that our strategic goods embargo is a leaky sieve that they not only import illegal military end use equipment, but ship it back to the West for **repair**, presumably secure in the belief that it can be reimported.

One example that was intercepted occurred in July 1977 when California Technology Corporation placed a purchase order with a U.S. manufacturer for $66,000 in components for sophisticated electronic machinery with direct military application. All components ordered were Munitions List items and cannot be legally exported without approval from the U.S. Department of State. Yet CTC received the equipment and in September 1977, under the name Interroga International Components and Sales Organization, CTC exported the components to West Germany.

Three years later one of the components was in need of repair. It was sent to the manfuacturer's plant for work. On June 16 and 23, 1981, in West Germany, Stephen Dodge of Customs, Robert Rice of Commerce, and Theodore W. Wu, Assistant U.S. Attorney in Los Angeles, received information that the machinery had been sold originally to Mashpriborintorg of Moscow, and the Russians had sent the disabled component back to ADT of Dusseldorf for repair.

A telex from CTC executives in Dusseldorf to Anatoli Maluta in Los Angeles, dated February 27, 1980, was seized by U.S. Customs agents. The telex said the component would be returned to the U.S. for repair. A "friend" would receive the repaired equipment and then turn it over to Maluta for re-export.

Now let's turn to the question of how the Soviets know **what** to order for their semiconductor plant. The invoices reproduced above suggest the Soviets knew precisely what production equipment they wanted to build a semiconductor plant. The question is how did they find out the model numbers, specifications and the rest?

CHAPTER V

Computers — Deception by Control Data Corporation

"We have offered to the Socialist countries only standard commercial computers and these offerings have been in full compliance with the export control and administrative directives of the Department of Commerce."

William C. Norris, Chairman
Control Data Corporation

To make any progress in developing weapons systems the Soviets must utilize modern high-speed computers. The computers and the necessary computer technology, both hardware and software, have come from the West, almost exclusively from the United States.

At the end of the 1950s the United States had about 5,000 computers in use, while the Soviet Union had only about 120. These Soviet computers, as reported by well-qualified observers, were technically well behind those of the West and barely out of the first-generation stage even as late as the 1960s.

In the late fifties the Soviets produced about thirty to forty BESM-type computers for research and development work on atomic energy and rockets and missiles. In general, the BESM type has most of the features typical of early U.S. computers. The original version had 7,000 tubes; the later version had 3,000 tubes and germanium diodes.

The only Soviet computer in continuous production in the 1960s was the URAL-I, followed by the URAL-II and URAL-IV modifications of the original model. The URAL-I has an average speed of 100 operations per second, compared to 2,500 operations per second on U.S. World War II machines and 15,000 operations per second for large U.S. machines of the 1950s, and 1-10 million operations per second common in the early 1970s. Occupying 40 square meters of floor space, URAL-I contains 800 tubes and 3,000 germanium diodes; its storage units include a magnetic drum of 1,024 cells and a magnetic tape of up to 40,000 cells — considerably less than U.S. machines of the 1960s. URAL-II and URAL-IV incorporate slightly improved characteristics. The URAL series is based on U.S. technology.

Production methods for both the URAL and the BESM computers were the same as American methods.

Until the mid-1960s **direct** import of computers from the United States was heavily restricted by export control regulations. In 1965 only $5,000 worth of electronic computers and parts were shipped from the United States to the Soviet Union, and only $2,000 worth in 1966. This changed in 1967. Computer exports increased to $1,079,000 and a higher rate of export of U.S. electronic computers to the USSR has been maintained to the present time under constant lobbying pressure from U.S. businessmen and their trade associations.

The precise amount and nature of U.S. computer sales to the Soviet Union since World War II is censored, but it is known that after World War II, IBM sales to the Communist world came "almost entirely from [IBM's] Western European plants," partly because of U.S. export control restrictions and partly because U.S. equipment operates on 60 cycles, whereas Russian and European equipment operates on 50 cycles.

American computer sales as opposed to Soviet theft may be traced from 1959 with sale of a Model-802 National-Elliott sold by Elliott Automation, Ltd., of the United Kingdom. (Elliott Automation is a subsidiary of General Electric in the United States.) Towards the end of the sixties Soviet purchases of computers were stepped up, and by late 1969 it was estimated that Western computer sales to all of Communist Europe, including the USSR, were running at $40 million annually, in great part from European subsidiaries of American companies. In 1964-65 Elliott Automation delivered five Model-503 computers to the USSR, including one for installation in the Moscow Academy of Sciences. Other General Electric made in Europe machines, for example, a Model-400 made in France by Compagnie des Machines Bull, were also sold to the USSR.

Olivetti-General Electric of Milan, Italy has been a major supplier of GE computers in the USSR. In 1967 the Olivetti firm delivered $2.4 million worth of data-processing equipment systems to the USSR in addition to Model-400 and Model-115 machines.

In sum, General Electric from 1959 onwards sold to the Soviet Union through its European subsidiaries a range of its medium-capacity computers.

Of perhaps even greater significance for the 1960 era were sales by English Electric, which include third-generation microcircuit computers utilizing Radio Corporation of America technology. In 1967 English Electric sold to the USSR its System Four machine with microcircuits; this machine incorporates RCA patent and was similar to the RCA Spectra-70 series.

The largest single supplier of computers to the USSR has been International Computers and Tabulation, Ltd. of the United Kingdom, which also licenses RCA technology, and by 1970 had supplied at least twenty-seven of the thirty-three large computers then in Russia. In November 1969, for example, five of the firm's 1900-series computers (valued at $12 million) went to the USSR. These large high-speed units with integrated circuits were, without question, considerably in advance of anything the Soviets were able to manufacture. Such machines were certainly capable of solving military and space problems. Indeed, a computer cannot distinguish between civilian and military problems.

In 1971 the USSR and East European family of general purpose computers known as the RYAD series was announced. These are based substantially on IBM 360 and 370 computers illegally diverted into the USSR. This had an important effect of making available to them a tremendous library of computer software that was RYAD compatible.

Dr. Baker has commented on the current RYAD position:

> In the area of available manpower, one of the serious problems afflicting the Soviet economy is the lack of qualified, highly trained, technical people in the areas of computers and microelectronics. One cause of this is the lack of enough computing and electronic equipment to train the next generation of scientists and engineers. They simply don't have enough equipment to allow students sufficient 'hands-on' practice at an early stage in their education. The Soviets are trying to alleviate this problem by producing large, for them, numbers of RYAD computers — copies of the U.S. IBM System 360's and 370's.[17]

Soviet Agatha — American Apple II

In mid 1983 the Soviets introduced their first personal computer — the AGATHA (a rather curious name for a Russian product).

Produced at Zelenograd, outside Moscow, the Agatha was reversed engineered from the APPLE II. Specifications are similar to the APPLE and the components are either Soviet produced from reverse engineered U.S. components or imported and bought openly or clandestinely in Europe or Japan.

Officials at the Academy of Pedagogical Sciences have admitted that the APPLE II served as a "prototype" for the Soviet Agatha.

In 1985 COCOM set up some new rules for microcomputers and made it legal to export without license low powered 8-bit computers. Such a machine sells in the West for $100 to $500.

The response was a flood of computer manufacturers attempting to make elaborate sales pitches to the multimillion Soviet microcomputer market. If history is any judge, the Soviets will buy a few thousand and then attempt to reverse engineer and produce in the Soviet Union. Presumably the microcomputers, although low powered, could have military applications and indeed this was openly admitted by a major computer manufacturer (*New York Times*, February 8, 1985):

"We have no illusions. Some of these are headed for the military."

[17]United States Senate, *Transfer of United States High Technology to the Soviet Union and Soviet Bloc Nations* Hearings before the Permanent Subcommittee on Investigations, 97th Congress Second Session, May 1982, Washington, D.C., p. 61.

Military End Use

Confirmation of military end use comes from unimpeachable sources — Soviet engineers who have worked on copying or reverse engineering in the Soviet Union and later defected to the West. These engineers have testified before Congress and provide firsthand evidence of Soviet military use of our technology. Here is a statement from Joseph Arkov, who graduated from a Soviet engineering school in 1970 and who now resides in the United States. Arkov worked in Soviet research installations.

> If, for example, a new American computer has been obtained by the Soviets, they will make a military application of it rather than a civilian application,"[18] and

> In my work in the second research installation I had the assignment of copying Western and Japanese high technology.

Arkov makes the interesting point that the Soviets are now so far behind technologically that they can no longer just reverse engineer as previously — they must import even the technology to manufacture high technology:

> They do not have the human resources or the fine tuned equipment required to produce the high technology machinery they try to copy. Once they know what makes a given piece of machinery work, they find that they do not have the technical know-how and equipment to produce the product themselves. That is why they want Western high technology machines that will enable them to produce the products. And the Western products they desire the most are those produced in the United States. That is why they want American high technology machines with which they can produce the components for high technology products.

Under Senate questioning Arkov confirmed that the major application of our high technology is for military end uses.

> **Mr. Arkov:** Well, the task of copying Western technology . . . part of their assignment was for military. Is that the question?

> **Senator Rudman:** Yes, let me just follow that up. You spoke in your prepared statement about the use of sophisticated American computers in various Soviet military operations, and also about the use of semiconductor technology. There are those in this country who feel that had we not transferred that technology legally to the Soviet Union — we sold them certain semiconductor technology and certain sophisticated computer technology in the

[18]United States Senate, *Transfer of United States High Technology to the Soviet Union and Soviet Bloc Nations* Hearings before the Permanent Subcommittee on Investigations, 97th Congress Second Session, May 1982, Washington, D.C., p. 27.

late sixties and early seventies — the Soviets would not have achieved the advantages in missilery which they have made in terms of the enormous throw weight and precision of their guidance systems. Do you agree with that assessment? Do you think that the sale of those semiconductors and those computers has given them a tremendous step forward in their technology in the defense area from your background and your knowledge?

Mr. Arkov: Yes, I think so. I can't tell exactly. It's hard to estimate the degree of advantage they got. But they gained there, using American computers and American semiconductors.[19]

Control Data Deception

The 1973 Control Data Corporation technical assistance agreement with the Soviet Union enabled the Soviets to complete phase one of their semiconductor manufacturing plant (see Chapter Four).

Highly significant is a comparison of Control Data Corporation's public argument to the media and Congress with this 1973 agreement and its totally one-sided presentation of the national security argument.

One can only conclude that some CDC statements are deliberate untruths. We make this statement by comparing Control Data public statements, particularly those of Chairman William Norris, with internal documents and agreements with the Soviet Union. These documents are confidential, but copies are in our possession.

On December 19, 1973 William Norris wrote Congressman Richard T. Hanna concerning public criticism of the CDC proposal to export advanced Cyber computers to the USSR.

We extract some statements from the letter (reproduced in full as Appendix C) and compare it to extracts from CDC internal documents reproduced on pages 61-67. For example,

> **Norris:** We have offered the Socialist countries only standard commercial computers and these offerings have been in full compliance with the export control and administrative directives of the Department of Commerce.

Comment: Reference to the 1973 Protocol of Intent between CDC and the USSR marked CONFIDENTIAL and reproduced here tells a vastly different story.

> **Norris:** Many persons including some of the witnesses before your Committee mistake the offering for sale of old or even current state of the art hardware for transfer of advanced technology. This is not unusual because in many cases it is difficult for those who are not technically well informed to distinguish advanced computer technology.

[19]Ibid.

Comment: Norris is comparing apples and oranges. What is "old" or "current" in the United States is far beyond "state of the art" in the Soviet Union. When Norris was offering a million operations per second CYBER computer to the Soviets, the run-of-the-mill Soviet technology was in the order of several thousands of operations per second, and that was on copies of imported equipment.

If multinational businessmen like William Norris were honestly mistaken in their information or somewhat shaky in their logic, then perhaps they could be forgiven. After all, to err is human.

Unfortunately, evidence proves beyond doubt that at least some of these deaf mute businessmen have deceived both Congress and the American public in an unseemly haste to make a buck.

We have documentary evidence in the case of Control Data Corporation and its Chairman, William Norris.

In the following pages we print a series of letters from Control Data Corporation to a concerned member of the public and contracts betweeen CDC and the Soviet Union:

* Exhibit A

Letter from American Council for World Freedom to its supporters identifying William Norris and CDC as exporters of valuable military technology to the Soviets.

* Exhibit B

Reply from William Norris, "Dear Yellow Card Sender," dated May 5, 1978. Note in particular the paragraph, "While we did sign an agreement for technological cooperation with the Soviet Union, we have **not** transferred any computer technology to them."

* Exhibit C

Full text of cooperation agreement cited by William Norris in Exhibit B.

* Exhibit D

"Letter (Protocol) of Intent" not mentioned by William Norris in Exhibit B, but which includes precise details of technology to be transferred, in distinct contrast to the Norris claim, "We have **not** transferred any computer technology to them."

Most importantly, it will be seen that Control Data Corporation transferred a vast range of information and technology to the Soviets, not only on computers but on manufacturing.

Publisher's Note: Due to the fact that some readers may have difficulty reading the reduced copies, we have reset the text in full in Appendix D.

Exhibit A — Letter from Fred Schlafly to friends and supporters of American Council for World Freedom, dated April 1978, asking to mail "Yellow Cards" of protest to William Norris.

Dear ACWF Supporter:

I need just a few seconds of your time right now.

I need you to sign and mail the 2 postcards I've enclosed for your personal use.

Since you're one of the best friends the American Council for World Freedom has, I'm sure I'm not asking too much of you.

Please let me explain these 2 postcards and tell you why it's crucial that you sign and mail them today.

The yellow postcard is addressed to Mr. William Norris, Chairman of the Control Data Corporation, one of the world's biggest and most advanced computer companies.

That yellow postcard calls on Mr. Norris and Control Data to stop selling computer technology to Communist Russia and its satellites...

...Technology that our Communist enemies are using to gain military superiority over the U.S.

Once before, Mr. Norris tried to sell our best and most advanced computer to the Soviet Union. Only an all-out, last minute effort by over 300 patriotic Congressmen stopped the sale of this highly valuable computer to the Russians.

Now Mr. Norris and Control Data are trying the same sell-out to Russia _again_. You and I have got to put a stop to it.

And just to make sure that Mr. Norris _personally_ sees _your_ postcard, I've addressed the card to his home address in St. Paul, Minnesota.

The blue postcard is addressed to Mr. J. Fred Bucy, President of Texas Instruments.

That blue postcard _praises_ Mr. Bucy and Texas Instruments for refusing to sell American technology

(over, please...)

to our Communist enemies.

You probably know from reading your anti-Communist magazines and newspapers that a lot of big, U.S. corporations are making huge profits selling their products to Communist Russia.

Now I'm not talking about selling products like soft drinks and clothing to the Communists.

I'm talking about U.S. companies...like Control Data...that sell our Communist enemies computers, shipbuilding equipment, and jet airplanes...

...Technology that Communist Russia is using to turn itself into the world's number one military superpower.

My friend, let me tell you just how critical a problem we really face.

There are over 60 big U.S. companies selling U.S. technological secrets to the Soviet Union...

...Companies like Union Carbide, General Electric, Armco Steel, Bryant Chucking Grinder, which sells the Communists the ball bearings they need for their attack missiles...and Control Data.

So why is ACWF only going after Control Data, and not these other companies, for selling out U.S. military superiority to the Communists?

The answer is,

> Control Data is absolutely, far and away, the
> biggest offender when it comes to selling out
> U.S. technology and actively aiding the Soviet
> military power grab.

There's no doubt in my mind that if we ACWF Supporters force Control Data to stop building the Soviet war machine, then other sell out companies like General Electric and Union Carbide will also stop aiding and abetting our Communist enemies.

That's why the American Council for World Freedom has set up its Task Force on Strategic Trade.

Look at what ACWF has already accomplished against Control Data:

* Just recently, we had a very large, enormously
  successful press conference on Capitol Hill,
  announcing our fight against Control Data's
  sell-out of U.S. computer secrets to the
  Communists. Over 75 reporters and Congressmen
  were there!

  (You might want to look at the enclosed note
  from General Daniel Graham which lists some of
  the Congressmen, newspapers, magazines, and
  wire services at our press conferenc.)

* ACWF just published a devastating expose of how Control Data and other sell-out companies are turning America into a second-rate power by helping build the Soviet war machine.

* We gave each reporter and Congressman at our press conference a copy of our study, "The Strategic Dimension of East-West Trade" by Dr. Miles Costick, one of America's top experts on foreign affairs and strategic trade.

ACWF is off to a fast start against Control Data, but there's no use kidding ourselves -- we've got a big fight ahead of us.

Here's what needs to be done right away:

1.) We've got to flood Control Data and its Chairman, Mr. William Norris, with postcards and letters demanding "No More U.S. Help for the Soviet War Machine."

You can help do that by signing and mailing your postcard to Mr. Norris...today.

2.) We've got to support and encourage companies like Texas Instruments when they show the guts and have the backbone to say "NO" to Communist Russia.

Will you help do that by mailing your postcard to Mr. J. Fred Bucy, President of Texas Instruments...today?

3.) We've got to print and distribute 28,959 copies of Dr. Costick's brilliant expose, "The Strategic Dimension of East-West Trade."

We must get this expose into the hands of America's 10,236 newspaper editors, 9,414 magazine editors, 9,309 TV and radio news editors.

4.) And finally, we've got to write to tens of thousands of other Americans...50 thousand in May alone... alerting them to Control Data and similar sell-outs of America's scientific secrets, getting them active in ACWF's fight to keep America the #1 military power on earth.

I know I don't have to tell you that I'm talking about an aggressive, expensive program.

It will cost $28,090 to print 28,959 copies of ACWF's Control Data expose and distribute them to 28,959 newspaper, magazine, radio and TV editors.

And it will cost $13,253 to write to fifty thousand Americans in May.

That's a total of $41,343 ACWF urgently needs. We simply don't

(over, please...)

Page 4

have it.

Will you help ACWF again, with as generous a gift as you can afford?

What I'm asking you for is your contribution...before May 15th... of at least $25, but hopefully more -- perhaps even $50, $100, $250, $500 or $1,000 if you can afford it.

I know I ask an awful lot of you, but I don't have anyone else to turn to.

So please, when you mail your 2 postcards, won't you use the enclosed reply envelope to rush ACWF your support of $25, $50, $100, $250, $500 or $1,000?

Gratefully,

Fred Schlafly

P.S. Remember, when you mail that yellow postcard protesting Control Data's military sell-out of the U.S. to Russia, your card will be read by Control Data Boss William Norris because it's addressed to his home.

Exhibit B — Letter from William C. Norris to each "Yellow Card Sender," dated May 5, 1978.

CONTROL DATA
CORPORATION

May 5, 1978

Dear Yellow Card Sender:

You are grossly unfair, woefully ignorant or being led around by the nose. I suspect it is the latter, otherwise you wouldn't have signed a card with a canned message, along with a request to contribute money to the author -- apparently a Mr. Schlafly of American Council for World Freedom.

Why didn't you write and ask about Control Data's position on trade with the Communists before signing a form card? You obviously have little knowledge about Control Data's position or activities, nor does Mr. Schlafly. He has not contacted anyone in Control Data. I would have furnished him information or I would have been glad to respond to a letter from you asking for information. Then if you still disagreed strongly -- so be it -- at least you would have acted sensibly and fairly by first getting adequate information. Then you probably wouldn't have resorted to the use of such words as "disgrace" and "outrage", because that is akin to the Communist procedure of not fairly weighing the evidence.

You obviously are deeply concerned about the Soviet military threat, and I certainly understand that, because I share that concern. I think about the subject a great deal and several points are clear, including that:

1. The U.S. should maintain a very strong military capability. Control Data's products and services contribute significantly to our nation's defense system.

2. The U.S. can gain much more by doing business with the Soviet Union than by trying to withhold many things. This latter procedure hasn't been effective where it has been tried.

3. Most important to maintaining a strong international position is that we maintain a strong domestic economy with an adequate number of jobs, especially for young, disadvantaged persons. (If you have studied world-wide Communism you are aware that it preys upon countries where the domestic economy is sick, weak or corrupt).

Further, Mr. Schlafly's letter to you was based on a highly
inaccurate booklet by Miles Costick. We are familiar with
Mr. Costick's work, and have met with him to document inaccuracies in it. Apparently, he has chosen to ignore this evidence,
because:

. While we did sign an agreement for technological
  cooperation with the Soviet Union, we have <u>not</u>
  transferred any computer technology to them.

. No Control Data computer is being used for any
  military purpose whatsoever by the Soviet Union.

. Further evidence of error was the blue card that
  you were asked to mail to Texas Instruments praising
  them for refusing to sell to the Soviets. I am
  sure that their chairman, Mr. Bucy, would admit that
  many Texas Instrument's high technology products are
  used in Soviet-built equipment.

My position is further explained in the enclosed article, "High
Technology Trade with the Communists", which appeared in
DATAMATION magazine last January.

I have also enclosed a clipping from a recent issue of TIME
magazine that describes some of Control Data's activities in
helping to provide more jobs for the disadvantaged.

If you have questions after reading the material, write me.

Sincerely,

William C. Norris

Encls.

Exhibit C — Letter (Protocol) of Intent dated 19 October 1973 (English version) between State Committee of the USSR Council of Ministers for Science and Technology and the Control Data Corporation.

LETTER (PROTOCOL) OF INTENT

I. The State Committee for Science and Technology of the USSR Council of Ministers and Control Data Corporation (CDC) in the interest of developing extensive long term technological relationships between Soviet organizations and enterprises and Control Data Corporation, have on 19 October 1973 concluded an "Agreement on Scientific and Technical Cooperation". The State Committee for Science and Technology and Control Data Corporation, together with the USSR Ministry of Foreign Trade, hereafter referred to as the Parties, having in mind furthering the development of the desired cooperation, wish to conclude this Letter of Intent to identify joint projects which need detailed study, and to commit the Parties to initiate these studies.

II. The objective of the projects called for in Paragraph III of this Letter (Protocol) of Intent shall be to:

A. Establish the coordinating group, and working groups called for in Article 4 of the Agreement on Scientific and Technical Cooperation.

B. Specify the scope and mechanism by which the scientific and technical information exchanges to be engaged in under the terms of Article 2 of the Agreement on Scientific and Technical Cooperation shall be carried out:

C. Implement detailed examination of those projects given first order of priority.

D. Define mutually satisfactory commercial and trade relationships.

III. The following preliminary list of tentative joint projects has been agreed to and a detailed examination of them will be our first priority.

A. Finance and Trade — To conduct studies and discussions in order to develop mutually satisfactory bases for payment for the activities to be undertaken within the scope of the Agreement on Scientific and Technical Cooperation. This shall include terms of credit, repayment

considerations, and cooperative activities whose
purpose is to generate Western currencies to facilitate repayment of credit.

B. Software and Applications Development - Development
by the USSR of automated programming methods for
improving programmer productivity as well as the
generation of systems within certain applications
areas of mutual interest. These would include, but
not be limited to, the medical/health care field,
the transportation industry, and the education environment.

Objectives of transportation study might include both
the increase of efficiency of passenger transportation,
as well as enhance the timely delivery of freight, both
by surface and by air.

The medical health care study would include the creation
of systems dealing with patient care, health care
management, as well as health care planning.

Work in the education field will be centered on the
disciplines of computer-based education and training,
including networks, special terminals, author language,
curriculum and courseware.

C. New Computer Sub-Systems R & D - The USSR will contribute
research and technical development of computer elements
based on new technical concepts. This research and
development will be undertaken based on cooperative
planning between Control Data Corporation and appropriate USSR organizations.

This research and development may include, but will not
be limited to, the realm of memory systems, advanced
mass memories, thin film, disk heads, plated disks,
memory organization and ion beam memories, might be
amongst the technologies to be investigated and research
and development undertaken.

D. Research and Development - Joint exploration of research
and development activities in USSR enterprises and institutes which will add to Control Data Corporation's
research and development activities, particularily in
the areas of computer systems software and applications
software, and will constitute in part compensation for
technologies which the USSR wishes to obtain from Control
Data Corporation.

E. Advanced Computer System Development - To conduct a cooperativ
development of a computer with rearrangeable structures and
a performance rate of 50 to 100 million instructions per
second to be used for solving large economic and management
problems. Results of software and architectural research
conducted in the Soviet Union will be utilized. The

functional design of the software system and the architectural concepts will be supplied by the USSR. The implementation would be conducted in four phases as follows:

1) Feasibility
2) Detailed design
3) Prototype construction and checkout
4) Test and evaluation

The feasibility phase will be used to evaluate the proposed computer concepts, both hardware and software, in order to verify that the system can be implemented with known hardware technology.

The following are among the areas to be considered as part of the evaluation: architectural design, software system design, new software techniques, new software applications, components and circuitry, system performance, operating systems, principal applications, manpower and cost. The entire program is expected to require from five to eight years to reach fruition. Results will be shared on an equal basis.

F. Disk Manufacturing Plant -- To build a plant for manufacturing mass storage devices based on removable magnetic disk packs with up to 100 million byte capacity per each pack. The yearly plant output shall be 5,000 device units and 60,000 units of magnetic disk packs (approximate estimate). It is expected that 80% of the plant output will be 30 mega-byte devices and 20% will be 100 mega-byte devices.

G. Printer Manufacturing -- To build a plant to manufacture line printers to operate at a speed of 1,200 lines per minute. The yearly output shall be 3,000 devices (approximate estimate).

H. Process Control Devices Manufacturing -- To build a plant for the manufacture of process control oriented peripheral devices, including data collection, analog/digital gear, terminals etc. The annual plant output for all devices, including data collection, is estimated as approximately 20,000 units.

Note: With respect to items F, G and H, the Parties shall jointly develop technological documentation for manufactured products in accordance with the metric system standard and insure that all aspects of the plant(s) design and construction are of the highest quality such that the manufactured output will be fully competitive in all

respects within the world market place. Control Data
Corporation shall deliver technical documentation,
complete sets of technological equipment, as well as
know-how, and full assistance in order to insure
mastering of full production techniques.

I. Printed Circuit Board Manufacturing - Delivery to the
USSR of a complete set of equipment for the manufacturing
of multilayer printed circuit boards.

J. RYAD/Control Data Corporation Information Processing Systems
Joint creation of information processing systems based on
the use of Soviet manufactured computers and Control Data
Corporation equipment. Control Data Corporation equipment
would include local peripheral sub-systems as well as
communications hardware, including front-ends, remote
concentrators and terminals. It is expected that the
system would include operational software jointly created
to make a viable computer system for applications to
problems in other parts of the world which could be
commercially saleable as a total system.

K. Process Control and Remote Communications Concentrator
Manufacture - To organize within the Soviet Union
manufacturing of Control Data Corporation licensed remote
communication equipment and analog to digital components
for standalone use within technological process control
systems. It must be kept in mind that such devices and
components must satisfy requirements of both the Soviet
Union and Control Data Corporation. Control Data Corporation evaluates that it can buy back approximately
$ 4,000,000 worth of these products.

L. Network Information System (CYBERNET) - Joint study to
provide a proposed specification for a Soviet computer
communication network. Dimensions of networks to be
explored will include both the distribution network
between terminal users and processing center as well as
an eventual bulk transfer network between cluster centers
and private satellite centers. The USSR will develop
systems and applications software for this Network Information System based upon the present commercially available
Control Data Corporation network software system. The
resultant software will be available and shared by both
Parties. The various data communications technologies will
be evaluated as one of the major tasks of the study.

M. Control Data CYBER 70 Data Processing Centers - Joint
creation of processing centers which will utilize computers
and other equipment of Soviet manufacture and these of

Control Data Corporation for organization/application such as, but not limited to, those specified below:

1) World Hydrometeorological Institute, Moscow, for processing of weather data for forecasting and in preparation of weather maps.

2) Institute of High Energy Physics, Serpukhov, for nuclear research, reduction of accelerated data, and basic particle research.

3) Ministry of Geology, Moscow, for seismic data reduction.

4) USSR Academy of Sciences, Academ Gorodok, Novosibirsk, for operational numerical weather forecasts, as well as for on-line concurrent processing of scientific data from experiments related to atmospheric, gas dynamics, weather research and basic scientific research.

5) Ministry of Chemical Industry, Moscow, for an information processing network and plant and pipeline control.

6) Ministry of Oil Refining, Moscow, for electric distribution and plant design.

IV. In order to launch the above noted programs into viable, successful implementations, Control Data Corporation is prepared to advance long term credit on mutually beneficial terms through Commercial Credit Company, its financial affiliate. It is expected that the total volume of credit required for financing of the full dimensions of the Scientific and Technological Agreement may reach 500 million dollars. It is further expected that repayment of such credit will be made in the form of Western currencies. Commercial Credit Company is prepared to assist the USSR in obtaining such currencies by working jointly with the appropriate authorities in the USSR in the worldwide marketing of Soviet products and in the development of natural resources for which sustained world demand exists.

In the process of developing specific programs to generate currency for credit repayment, Control Data Corporation offered the following proposals for consideration:

A. Financing -- Control Data Corporation is prepared to assist the USSR in obtaining the financing needed to accomplish the above noted objectives. Through its wholly-owned subsidiary, Commercial Credit Company, Control Data Corporation can marshal substantial long term credit to facilitate Soviet purchases of equipment and technology projected under the Agreement, as well as credit for potential customers for worldwide sales of Soviet products and materials.

Commercial Credit Company will take the initiative in assembling consortiums of major lenders to extend long term credit to the USSR on mutually beneficial terms. Commercial Credit Company is also prepared to use its Luxembourg facility to make use of U.S. Export/Import Bank financing to the extent such credit is available to finance the export of Control Data Corporation equipment to the USSR.

B. Joint Venture Marketing Company -- A marketing company jointly owned by a competent Soviet organization and Commercial Credit Company will be established. The company would be incorporated in Western Europe and would be a self-sufficient, profit oriented entity.

The prime purpose of the company would be to sell and service non-computer-related Soviet products in the Western world on the open market. In addition to selling the products of the USSR the Joint Venture Marketing Company will develop, in conjunction with Commercial Credit Company, economic and trade forecasts for the near to medium term future in the Western world that will indicate market trends, requirements and shortages, which can be fed back into the USSR five-year planning activities to insure that future USSR exports meet the requirements of future Western markets.

C. Natural Resource Development - Soviet natural resources are the largest of any single country in the world. The key to world trade in excess natural resources, not required for domestic use, is the creation of additional facilities to obtain and profitably dispose of these resources.

The financial resources and worldwide industry associations of Commercial Credit Company will be employed to create appropriate consortia having four major functional entities:

a. finance
b. resource extraction and processing expertise
c. construction and operation expertise
d. sales outlet for processed products

Natural resources to be considered would include, but not be restricted to: timber products, non-ferrous metals, precious metals and stones, asbestos, apatite, cement, synthetic rubber and resins, and coal.

The present Letter of Intent is executed in two versions, both in Russian and English. Both versions of the text have equal validity.

For the State Committee
of the USSR Council of
Ministers for Science
and Technology

For Control Data
Corporation

For the USSR Ministry
of Foreign Trade

Exhibit D — English version of Agreement between State Committee of the Council of Ministers of the USSR for Science and Technology and Control Data Corporation (signed by Robert D. Schmidt), dated 19 October 1973.

on Scientific and Technical Cooperation
between the State Committee of the Council
of Ministers of the USSR for Science and Technology
and Control Data Corporation(USA)

The State Committee of the Council of Ministers of the USSR
for Science and Technology(GKNT) and Control Data Corporation
(CDC), hereinafter called "Parties",

Considering that favourable conditions have been created
for extensive development of a long-term scientific
and industrial and economic cooperation;

Taking into account the mutual interest of both Parties
in the development of such cooperation and recognizing
the mutual advantage thereof; and

In accordance with Paragraph 8 of the "Basic Principles
of Relations between the Union of Soviet Socialist
Republics and the United States of America", signed
on May 29,1972, and Article 4 of the "Agreement Between
the Government of the USSR and the Government of the USA
on Cooperation in the Fields of Science and Technology"
concluded on May 24, 1972;

Have agreed as follows:

ARTICLE 1

The subject of the present agreement has to do with
a long-term program for a broad scientific and technical
cooperation in the area computational technology, and
specifically;

-To conduct joint development of a technically
advanced computer;

-Joint development and organization of the production
of computer peripheral equipment;

-Joint creation of information processing systems
based on the technical means of Soviet production
and on the technical means developed by CDC and
the development of software means for these systems;

-Joint development of Analog to Digital Equipment for control systems of technological processes;

-Joint development of computer components, technical equipment for their production and the organization of production of these components.

--Development of computer memories (based on large volume removable magnetic disk packs, and on integrated circuits, etc.).

-Creation of equipment and systems for data communication;

-Application(use) of computers in the fields of medicine, education, meteorological, physics, and etc.;

-Preparation (training) of specialists in the area of computer technology;

The scope of this Agreement may at any time be extended to include other fields of specific subjects of cooperation by agreement of the Parties.

This Agreement is not limiting either Party from entering into similar cooperation in the said fields with a third Party.

ARTICLE 2

Scientific and technical cooperation between the Parties can be implemented in the following forms with specific arrangements being exclusively subject to mutual agreement between appropriate Soviet organizations and the firm of Control Data Corporation:

-Exchange of scientific and technical information, documentation and production samples;

-Exchange of delegations of specialists and trainees;

-Organization of lectures, symposia and demonstrations of the production samples;

-Joint research, development and testing, exchange of research results and experience;

-69-

-Mutual consultations for the purpose of discussing and analysing scientific and technical problems, technical principles, ideas and concepts in the appropriate areas of cooperation; .

-Creation of temporary joint research groups to perform specific projects and to produce appropriate(joint) reports.

-Exchange, acquisition or transfer of methods, processes, technical equipment, as well as of "know-how" and of licenses for the manufacture of products.

ARTICLE 3

The Parties have established that financial, commercial, and legal questions related to advancement of credit and payments for the delivered products and technical equipment, assignation of licences and "know-how" as well as supplied services in performance of the various joint projects, relative to the present Agreement, shall be decided by separate agreements between appropriate competent. Soviet organizations and the Control Data Corporation.

ARTICLE 4

For the practical implementation of the present Agreement the Parties shall establish a Coordinating group, from authorized representatives (coordinators) which shall determine and recommend a proper course for the cooperation and also to control compliance with responsibilities assumed by the Parties, and to take the necessary action for the successful implementation of the objectives of the present Agreement. For the preparation of proposals for the concrete cooperative projects, there shall be established special groups of experts whose task it will be to determine technical and economic feasibility of the joint projects and to draw up action plans for their realization. The results of these working groups shall be turned over to the Coordinating group for their discussion and preparation of recommendations.

Recommendations and proposals of the Coordinating group will be presented in the form of protocols, which will be used as the basis for preparation of separate protocols or contracts.

Coordinating and working groups shall meet as frequently as is necessary to perform their functions alternatively in the USSR and USA unless otherwise agreed.

ARTICLE 5

Scientific and technical information furnished by one Party
to the other under this agreement may be used freely for its
own research, development and production, as well as the
realization of finished products unless the Party supplying
such information stipulates at the time of its transfer that
the information may be used only on the basis of special
agreement between Parties. This information can be transmitted to a third Party only with the approval of the Party
which has furnished it.

Information received from a third Party which cannot be disposed of at will by one of the Contracting Parties is not
subject to transmittal to the other Party unless mutually
satisfactory arrangements can be made with the third Party
for communication of such information.

It is contemplated in the foregoing that any organizations
or enterprises of the USSR and any wholly owned or partially
owned Control Data subsidiaries shall be not regarded as a
third Party.

ARTICLE 6

Expenses of travelling back and forth of specialists of both
Parties under the programs related to this Agreement, as a
rule will be defrayed as follows:

   -The Party sending the specialists pays the round-trip
   fare.

   -The host Party bears all costs connected with their
   stay while in its own country.

The duration of the above visits and the number of specialists
in each group shall be mutually agreed to by the Parties in
advance of the visits.

Organizational questions, arising from implementation of this
present Agreement shall be discussed and determined by the
Parties in the course of working.

The present Agreement shall continue for a period of 10 (ten) years and shall enter into force immediately upon its signature. It can be extended with mutial agreement of the Parties.

The cancellation of the present Agreement shall not affect the validity of any agreement and contracts enctered into in accordance with Article 3 of the present Agreement by organizations and enterprizes of the USSR and CDC.

Drawn up and signed the *19 October* 1973, in the city of Moscow, USSR, in duplicate, one copy in Russian and one in English, both texts being equally authentic.

For the State Committee of the
Council of Ministers of the USSR
for Science and Technology

For the Control Data
Corporation

Robert D Schmidt

The Deceptive World View of Control Data Corporation

William Norris, Chairman of Control Data, has a lively correspondence with Americans anxious to learn his rationale for supporting the Soviet Union.

We quote below an extract from a letter written by William Norris to an inquirer:

You also made reference [wrote Norris] in your letter to Russia's first democratic government that was overthrown by the communists. You are incorrect on this point. There never has been any democracy in Russia — as a matter of fact, the Russian standard of living today is higher than it was under the tsars. Further, you don't find a great deal of unhappiness in the Soviet Union over living conditions and the communist regime for two reasons — (1) they have never know [sic] what democracy is, and (2) life is better than it used to be.

Here are the errors in the above Norris paragraph:

• "There has never been any democracy in Russia."

Incorrect. The Kerensky government from March to November, 1917 was freely elected and overthrown by the Bolsheviks (with the aid of Western businessmen such as William Norris).

• "You don't find a great deal of unhappiness in the Soviet Union over living conditions."

Incorrect. Mr. Norris should look out the window of his Moscow office at the uniform drab blocks of apartments. How many families live in one room? How often do several families live in one apartment? How about the hours spent in food lines, and the limited choice of consumer goods in a guns-before-butter economy? Just how many individual Russians has Mr. Norris freely spoken with? Not those of the "nomenklatura," but average Russians in the street. We venture to guess none at all.

• ". . . standard of living today is higher than under the tsars."

Take one item — wheat. In 1906 Russia was the world's largest wheat exporter and the world's largest wheat producer. The climate is the same today as in 1906, yet is used as a forlorn excuse for Soviet pitiful wheat production. In fact, 80% of Russian bread today is made from **imported** wheat, the home-grown is only fit for cattle feed. Without Western wheat, Russia today would starve. Is that a truly higher standard of living? Anyway, Russians today don't compare their standards to those of tsarist times but to the Western world.

William Norris only sees what he wants to see, hears what he wants to hear, and presumably speaks from these limited impressions of the world.

In conclusion, we can thank Mr. Norris and Control Data Corporation that Soviet military has been able to break into the electronics based warfare of the late 20th and early 21st century.

CDC fulfilled phase one of the Soviet program for acquisition of Western semiconductor technology and mass production facilities.

Chapter VI

Soviets in the Air

Before we got the (U.S.) guidance systems we could hardly find Washington with our missiles. Afterwards we could find the White House.

Without U.S. help the Soviet military system would collapse in 1½ years.

Avraham Shifrin, former Soviet Defense
Ministry official

Signal rockets were used in the Russian Tsarist Army as early as 1717. Present Russian theoretical work in rockets, beginning in 1903, stems from K. E. Tsiolkovskii, who investigated atmospheric resistance, rocket motion, and similar problems. This Tsarist work was continued in the Soviet Union during the twenties and thirties. In 1928 pioneer Tsiolkovskii suggested that the value of his contribution had been in theoretical calculations. Nothing had been achieved in practical rocket engineering.

Then in 1936, V. F. Glushke designed and built a prototype rocket engine, the ORM-65. This rocket used nitric acid and kerosene as propellants. The Soviets then developed the ZhRD R-3395, an aircraft jato rocket using nitric acid and aniline as a propellant. Du Pont provided technical assistance and equipment for the construction of large nitric acid plants. During World War II, Soviet rockets used "Russian cordite," which was 56.5 percent nitrocellulose. The nitrocellulose was manufactured under a technical-assistance agreement made in 1930 with the Hercules Powder Company of the United States.

Finally, under Lend-Lease, 3,000 rocket-launchers and large quantities of propellants were shipped from the United States to the USSR.

German Assistance for Soviet Rockets

A major boost to Soviet ambitions in rocketry came from Germany at the end of World War II. Facilities transferred to the USSR included the rocket testing stations of Blizna and Peenemunde, captured intact and removed to the USSR; the extensive production facilities for the V-1 and V-2 at Nordhausen and Prague; the records of reliability tests on some 6,000 German V-2; and 6,000 German technicians (not the top theoretical men), most of whom were not released from Russia until the late 1950s.

The German rocket program was in an advanced state of development in 1945. About 32,050 V-1 "Flying bomb" weapons had been produced in the Volkswagen plant at Fallersleben and in the underground Central Works at Nordhausen. In addition, 6,900 V-2 rockets had been produced — 6,400 at the underground Mittelwerke at Nordhausen and 500 at Peenemunde. Rocket fuel facilities had been developed in the Soviet Zone: liquid oxygen plants at Schmeidebach in Thuringia and at Nordhausen, and a hydrogan peroxide plant at Peenemunde.

The Germans undertook two and one-half years of experimental work and statistical flight and reliability evaluation on the V-2 before the end of the war. There were 264 developmental launchings at Peenemunde alone.

Mittelwerke at Nordhausen was visited in June 1946 by U.S. Strategic Bombing Survey teams who reported that the enormous underground plant could manufacture V-1s and V-2s as well as Junkers-87 bombers. V-2 rockets were manufactured in twenty-seven underground tunnels. The plant was well equipped with machine tools and with "a very well set up assembly line for the rocket power unit." Its output at the end of the war was about 400 V-2s per month, and its potential output was projected at 900-1,000 per month.

When the Soviets occupied part of the American Zone in July 1945 under arrangement with General (later President) Eisenhower, the Nordhausen plant was removed completely to the USSR.

The United States and Britain never did gain access to German rocket-testing sites in Poland. The Sanders Mission reached the Blizna test station, after considerable delays in Moscow, only to find that its equipment had been removed "in such a methodical way as to suggest strongly to the mission's leader that the evacuation was made with a view to the equipment being reerected elsewhere." The Sanders Mission accumulated 1.5 tons of rocket parts. Unfortunately, when the mission reached home it found that the rocket parts had been intercepted by the Soviets. Rocket specimens so carefully crated in Blizna for shipment to London and the United States were last seen in Moscow. The crates arrived at the Air Ministry in London, but contained several tons of "old and highly familiar aircraft parts when they were opened." The Blizna rocket specimens had vanished.

Many German rocket technicians went or were taken to the Soviet Union. The most senior was Helmut Groettrup, who had been an aide to the director of electronics at Peenemunde. Two hundred other former Peenemunde technicians are reported to have been transferred. Among those were Waldemar Wolf, chief of ballistics for Krupp; engineer Peter Lertes; and Hans Hock, an Austrian specialist in computers. Most of these persons went in the October 22-23 round-up of ninety-two trainloads comprising 6,000 German specialists and 20,000 members of their families. Askania technicians, specialists in rocket-tracking devices, and electronics people from Lorenz, Siemens, and

Telefunken were among the deportees, as were experts from the Walter Rakententriebwerke in Prague.

Asher Lee sums up the transfer of German rocket technology:

> The whole range of Luftwaffe and German Army radio-guided missiles and equipment fell into Russian hands. There were the two Henschel radar-guided bombs, the Hs-293 and the larger FX-1400 . . the U.S.S.R. also acquired samples of German anti-aircraft radio-guided missiles like the X-4, the Hs-298 air-to-air projectile with a range of about a mile and a half, the Rheintochter which was fitted with a radar proximity fuse, and the very promising Schmetterling which even in 1945 had an operational ceiling of over 45,000 feet and a planned radius of action of about twenty miles. It could be ground- or air-launched and was one of the most advanced of the German small-calibre radio-guided defensive rockets; of these various projectiles the Henschel-293 bomb and the defensive Schmetterling and Hs-298 [the V-3] are undergoing development at Omsk and Irkutsk . . . [and later at] factories near Riga, Leningrad, Kiev, Khaborovsk, Voronezh, and elsewhere.

Other plants produced improved radars based on the Wurzberg system; the airborne Lichenstein and Naxos systems were reported in large-scale production in the 1950s.

The Soviets froze rocket design in the late 1950s on developments based on German ideas. The German technical specialists were sent home. By 1959 the Soviets landed a rocket on the moon.

Sputnik, Lunik and the Soyuz Programs

From the German V-2 rockets, associated German production facilities, and the all-important German reliability tests, stem the contemporary Soviet ICBM and space rockets.

In the 1960s there were four types of large liquid rockets in the Soviet Union: the Soviet version of the V-2, the R-10 (a 77,000 pound thrust scale-up of the German V-2), the R-14 (a scaled up V-2 with 220,000 pound thrust), and a modification known as R-14A (based on the R-14). The R-14 was designed and developed by a joint German-Russian team. The Germans were sent home in the late 1950s.

The Soviets did not until fairly recently use single boosters — they use clusters of rockets strapped onto a central core. The strap-ons were the scaled-up and modified German V-2. Thus, for example, **Sputnik I** and **Sputnik II** had a first stage of two R-14A units, a second stage of two R-14A units, and a third stage of a single R-10 (the German V-2 produced in the Soviet Union). **Lunik** was a similar cluster of six rocket units. The **Vostok** and **Polyot** series are clusters of six units. The planetary rockets, **Cosmos** series and **Soyuz** family are seven-unit clusters. An excellent photograph of one of these cluster vehicles is to be found in Leonid Vladimirov's book, *The Russian Space Bluff*.[20]

In the mid-sixties, any foolhardy person who insisted that the United States would be first on the moon because the Russians were technologically backward was dismissed as a dimwitted neanderthal. But at least two skilled observers with firsthand access to the Soviet program made a detailed case, one in 1958 and one in 1969. Lloyd Mallan wrote *Russia and the Big Red Lie* in 1958, after an almost unrestricted 14,000 mile trip through Russia to visit thirty-eight Soviet scientists. He took 6,000 photographs. It was Mallan who first drew attention to the Soviet practice of illustrating space-program press releases with photographs from the American trade and scientific press. The Remington Rand Univac computer was used in the fifties to illustrate an article in *Red Star* on the Soviet computer program — the captions were translated into Russian. In 1969, *Tass* issued a photograph for use in American newspapers purporting to show a Russian space station at the time when one Soviet space ship was in orbit and another en route. The *Tass* photograph was reproduced from *Scientific American* (Feb. 1962) and was identical to an advertisement in Sperry Gyroscope Company of Great Neck, New York. Sperry commented, "Apparently it is the same as the ad we ran."

This author personally remembers an incident from the early 1960s which illustrates the extraordinary success of Soviet propaganda in molding the U.S. concept of Russian technology. After giving a short speech to a Los Angeles audience on technological transfers to the Soviets, a member of the audience asked a sensible question: "Who will be first on the moon, the U.S. or the Soviets?"

The answer as closely as can be recalled was to the effect that the Soviets did not have the technology to be first on the moon, and by themselves could not make it in this century.

The response from the audience was an instantaneous and loud laugh — how ridiculous was the general audience response, "everyone knows" the Soviets are far ahead of the United States in space.

In fact, the Soviets could not even have achieved their Soyuz program without U.S. help. The docking mechanism is a direct copy of the U.S. docking mechanism.

Unfortunately, NASA and U.S. planners have a conflict of interest. If they publish what they know about the backwardness and dependency of the Soviet space program, it reduces the urgency in **our** program. This urgency is vital to get Congressional funds. Without transfers of technology the U.S. is in effect racing with itself, not a very appealing argument to place before Congress.

[20](London: Tom Stacey, Ltd., 1971), p. 88.

Why Did the Soviets Embark on a Space Program?

From an economic viewpoint, a Soviet space program makes no sense at all: such a program only makes sense from a geopolitical viewpoint.

In 1957, the year of **Sputnik**, the Soviet Union had fewer telephones than Japan (3.3 million in the USSR versus 3.7 million in Japan). On a per-hundred population basis, the Soviet Union could provide only 3.58 telephones compared to 49.8 in the United States. Even Spain provided 9.6 telephones per 100 of population, or three times more than the Soviet Union.

In automobiles, the Soviet Union was even less affluent. In 1964 the Soviet Union had a stock of 919,000 automobiles, all produced in Western-built plants, only slightly more than Argentina (800,000) and far less than Japan (1.6 million) and the United States (71.9 million).

Even today the Soviet Union is so backward in automobile technology that is has to go to Italy and the United States for automobile and truck technology.

Although **we** in the West might see this technical backwardness as a natural reason for **not** going into space, the Soviets saw it as a compelling reason to embark on a space program.

A "technical extravaganza" was necessary to demonstrate Soviet "technical superiority" to the world and maintain the myth of self-generated Soviet military might.

The Soviet economic problem in the mid-1950s was acute. The Soviet economy had shown good rates of growth, but this was due to the impetus given by Lend-Lease equipment and by war reparations. There were no signs of technical viability. Numerous industries were decades out of date with no indigenous progress on the horizon. The only solution was a massive program of acquiring complete plants and up-to-date technology in the West. Beginning in the late 1950s and continuing through to the 1980s, this program had to be disguised because of obvious military implications. One facet of the disguise was the space program. The usual stock of reasons for backwardness had run dry (the Civil War, the Revolution, intervention, warmongering capitalists) — even the damage done by the Nazis could only be spread so far. So two new elements made their appearance:

1. A space program — to get the Western world looking upwards and outwards, literally away from the Soviet Union and its internal problems.

2. Concurrent articles and press releases in the West on Soviet technical "achievements," spotted particularly in Western trade journals and more naive newspapers, such as the *New York Times*.

Around the same time in the '60s and '70s, the West (or rather the United States and Germany) resurrected Edwin Gay's 1918 proposal to mellow the Bolsheviks, and this proposal now became "bridges for peace" to provide a rational explanation for the massive transfers of Western technology that were required to fulfill Soviet programs. The United States appears, in historical perspective, to have been almost desperate in its attempts to help the Soviets in space. Of course, if the Soviets did not succeed in space, there could be no "competing" American space program and many politicians, bureaucrats, and politically oriented scientists were determined — for their own good reasons — that there had to be a major American space effort. There also had to be U.S. assistance for the Soviet space program.

In the ten years between December 1959 and December 1969, the United States made eighteen approaches to the USSR for space "cooperations."

In December 1959, NASA Administrator R. Keith Glennan offered assistance in tracking Soviet manned flights. On March 7, 1962 President Kennedy proposed an exchange of information from tracking and data-acquisition stations, and on September 20, 1963 President Kennery proposed joint exploration of the moon, an offer later repeated by President Johnson. On December 8, 1964 the administration proposed an exchange of teams to visit deep-space tracking and data-acquisition facilities. On May 3, 1965 NASA suggested joint communications tests via the Soviet **Molniya I.** On August 25, 1965 NASA asked the Soviet Academy of Sciences to send a representative to the **Gemini VI** launch, and on November 16, NASA inquired about joint **Molniya I** communications tests. Four U.S. offers were made in 1966; in January NASA inquired about Venus probes; on March 24, and May 23 Administrator James Webb suggested that the Soviets propose subjects for discussion; and in September Ambassador Arthur Goldberg again raised the question of tracking coverage by the United States for Soviet missiles.

Soviet Aircraft Development

In 1913 in St. Petersburg, Igor Sikorsky (who later founded the Sikorsky Aircraft Company in America) designed the "Russki-Vityazyi." Weighing 5 tons with a load of seven passengers, this four-engined plane established a contemporary endurance record of 1 hour and 54 minutes aloft. By 1917 a fleet of seventy-five IM ("Ilya Mouremetz") four-engined bombers, based on the original 1913 model, were in service — several decades before the American four-engined bomber fleets of World War II. So tsarist Russia produced and successfully flew the world's first four-engined bomber, a quarter of a century before the United States developed one. This early bomber had a wing span of over 100 feet, or only 21 inches less than that of the World War II Boeing B-17 Flying Fortress.

Obviously there was nothing wrong with indigenous Russian aeronautical talent half a century ago. While Russians have a natural affinity and geographic impulse towards aeronautics, the Soviets have only kept up with the West by reverse engineering, prolific "borrowing" and importation of technology and manufacturing equipment. Russian dependency on Western aeronautical design and production equipment and techniques goes back to the early 1920s.

At that time, soon after the Bolshevik Revolution, the Russian aircraft industry depended heavily on foreign aircraft and engine imports. There was considerable Soviet design activity, but this work was not converted into usable aircraft technology. Consequently, in the early 1930s the Soviet stock of military planes was almost completely foreign: 260 fighters comprising 160 De Havilland Type 9a (from Great Britain) and 100 Heinkel HD-43 fighters (from Germany); 80 Avre 504-K training biplanes (from Great Britain) and some Moraine-Saulnier monoplanes (from France); 52 R-3 biplanes (Russian TsAGI design); 20 R-6 reconnaissance planes (Russian TsAGI design); 242 I-4 Jupiter-engine planes (from Great Britain); 80 Ju-30 and ANT-6 (Junkers design); 20 ANT-6 bomber seaplanes (Russian design; 18 Avro-504L seaplanes (from Great Britain); 40 Savoia S-62 scouting flying boats (from Italy); 150 Heinkel HD-55 scouting flying boats (from Germany); 46 MR-4 (Savoia S-62 license) flying boats (from Italy); 12 TBI (Russian TsAGI design); and 43 Ju-30 naval bombers (from Germany).

From about 1932 onward, and particularly after 1936, there was extensive acquisition of Western aeronautical advances, which were then integrated with the developments of the 1920s. Fortuitously for the Soviet Union, this much-needed acquisition coincided with a period of increased competition among Western aircraft manufacturers. In many cases, military aircraft were designed in the West on Soviet account and the heavy, slow, underpowered Russian designs of the early 1930s were replaced by efficient Western designs.

By 1937 the Soviet government was convinced that the American method of building aircraft was the best for Russian conditions, as the American system of manufacture could more easily be adapted to mass production than any of the European systems. The United States thus became the main source of Soviet aircraft technology, particularly as a builder of new Soviet aircraft plants. Between 1932 and 1940 more than twenty American companies supplied either aircraft, accessories, or technical assistance for complete planes and aircraft manufacturing plants. Technical assistance agreements were made for Vultee attack bombers, the Consolidated Catalina, the Martin Ocean flying boat and Martin bombers, Republic and Sikorsky amphibians, Seversky amphibians and heavy bombers, Douglas DC-2 and DC-3 transports, the Douglas flying boat, and other aircraft. Kilmarx has well summarized this acquisition:

> The objectives of the Soviet Union were more straightforward than its methods. By monitoring aeronautical progress and taking advantage of commercial practices and lax security standards in the West, the Russians sought to acquire advanced equipment, designs, and processes on a selective basis. Emphasis was placed on the legitimate procurement of aircraft, engines (including superchargers), propellers, navigational equipment, and armament; specifications and performance data; design, production and test information and methods; machine tools, jigs and dies; semi-fabricates and critical raw materials. Licenses were obtained to manufacture certain modern military aircraft and engines in the U.S.S.R. At the same time, a number of Soviet scientists and engineers were educated at the best technical institutes in the West. Soviet techniques also included assigning purchasing missions abroad, placing inspectors and trainees in foreign factories, and contracting for the services of foreign engineers, technicians and consultants in Soviet plants.[21]

In 1937 the Soviet Union possessed the world's first commercial plane able to fly the Atlantic Ocean nonstop, with a payload of 7,500 pounds. Known as the Martin Ocean Transport, Model-156, with four 1,000 horsepower Wright Cyclone engines, it was built by the Glenn L. Martin Company of Baltimore. Model-156 cost the Soviet Union $1 million. Although capable of being flown directly to the Soviet Union, it was flown only to New York, then was dismantled and shipped to the USSR by boat.

Also in 1937 the Martin Company agreed to design a Soviet bomber. Loy Henderson, the U.S. charge in Moscow, reported:

> . . . since January 1, 1937, the Embassy granted visas to fourteen Soviet engineers and specialists who are proceeding to Baltimore to the Glenn L. Martin factory. This information would appear to be significant in view of the statements that the Martin Company is to design and develop a new type of large plane for the Soviet air force instead of selling somewhat obsolete models which may have been released for export by the American military authorities . . .[22]

In May 1937 the *New York Times* reported a $780,000 Soviet contract with Seversky Aircraft Corporation involving construction of, and

[21]R.A. Kilmarx, *A History of Soviet Air Power* (New York: Praeger, 1962).

[22]U.S. State Dept. Decimal File, 711.00111 — Armanent Control/1384, Nov. 4, 1938.

manufacturing rights for, Seversky amphibians, which then held the amphibian world speed record of 230.4 miles per hour. Under a technical-assistance agreement, Seversky Aircraft provided assistance for manufacture of these planes in the Soviet Union at the rate of ten per day.

Alexander P. de Seversky, president of the company, then informed the State Department that the Soviets had contracted to purchase from the company a large number of bombing planes of a new type to be designed by him. After being informed that a license would be granted if the planes involved no military secrets, Seversky suggested that the War and Navy Departments might object to its exportation "merely" on the ground that the new bomber would be superior to any bombing plane then in existence. Seversky indicated that he intended to address his request for an export license to the State Department, "in hope that the Department might expedite action in this."

The first domestic flying boats under the Soviets were constructed at Leningrad and Taganrog. In 1932, Plant No. 23 in Leningrad produced 18 Avro 504-L seaplanes and 40 Savoia S-62 scouting flying boats, the latter under a license from the Societa Idrovolanti Alta Italia of Milan — an outstanding designer of high-performance flying boats. Also in 1932, Taganrog Plant No. 31 produced 196 flying boats: 150 scouting HD-55s and 46 MR-5s, both built under license from Heinkel. The Soviets also acquired a license from the Macchi Company of Italy to produce the MBR series of Russian flying boats.

Then in 1937 came an agreement with the Consolidated Aircraft Company of San Diego for technical assistance for Catalina flying boats under supervision of Etienne Dormoy. With the Catalina flying boat we once again see the extraordinary ability of the Soviets to acquire anything they set their hearts on. The very **first** commercial Consolidated PBY ("Catalina") off the assembly line in San Diego was sold to the American Museum of Natural History — which promptly transferred it to the Soviets.[23] This is not the first time the American Museum of Natural History turns up in the Soviet files. In 1919 a shipload of Soviet propaganda was seized — en route to the United States and addressed to the American Museum of Natural History.[24]

Also in 1937-38, the Vultee Aircraft Division of Aviation Manufacturing Corporation of Downey, California built a fighter aircraft plant for the Soviets in Moscow.

Equally as important, the Soviets acquired rights to build the famous Douglas DC-3, probably the most successful transport plane in the history of aviation. Donald Douglas produced his first DC-3 in March

[23]*Aircraft Year Book*, 1938, p. 275.
[24]U.S. State Dept. Decimal File, 316-25-684.

1935 and within one year the Soviets decided this was to be the basic transport plane for the USSR. A technical-assistance agreement with the Douglas Aircraft Company was signed on July 15, 1936 for three years. Within thirty days of contract signature, Douglas delivered the blueprint materials required to fulfill the assistance contract.

In October 1937 the Soviet aircraft industry placed a $1.15 million order with Douglas for additional parts, tools, assemblies, and materials. The order included one complete DC-3 in subassembly and another in "first-stage" production. In addition, aluminum extrusions were ordered for another fifty aircraft, together with two complete sets of raw materials and twenty-five sets of finishing materials ranging from ash trays to zippers. Construction facilities, ordered at the same time, included one complete set of 6,485 templates, a set of 350 lead and zinc drop hammer dies, three sets of hydraulic mechanisms, all the necessary wood and plaster patterns, drill and assembly fixtures, a complete set of drop-hammer stamps, hydraulic-press parts, two crowning machines, and a set of 125 special tools. Later, another six complete transports were purchased, but it was not until 1940, four years after the agreement, that the Soviets got any domestic DC-3s (renamed the PS-84 or the LI-2) off a Soviet assembly line.

For input materials for military aircraft operation and construction the Soviets also depended on American construction assistance and technology. Even after the extensive American construction of refineries in the early 1930s the Soviet Union continued to be dependent on American technology for cracking petroleum into light gasoline fractions. Lend-Lease equipment deliveries brought the output of aviation gasoline from a mere 110,000 metric tons per year in 1941 to 1.65 million metric tons in 1944, despite the fact that several Lend-Lease cracking units were not delivered until after the end of the war. The Standard Oil Company of New York supplied the Soviet Union with technical information, plant designs, and a pilot manufacturing plant for sulfuric acid alkylation for production of 100-octane gasoline, and "voltolization" of fatty oils for production of aviation lubricating oils.

Efficient and specialized tools were developed by American aircraft manufacturers and their equipment suppliers and these in turn were purchased by the Soviets. For example, in 1938 the Lake Erie Engineering Corporation received a Soviet order for six hydraulic presses for forming metal aircraft sections. In the same year, Birdsboro Steel Foundry and Machine Company of Birdsboro, Pennsylvania, filled a half-million-dollar order for hydraulic presses for aircraft manufacture. Similarly, in 1938 the Wallace Supplies Manufacturing Company of Chicago, Illinois, sold seven bending machines "specially designed to bend tubing for aircraft and parts of motors" for $34,000. Most, if not all, Soviet aircraft accessories were straight copies of foreign products. When biplanes were used, "the streamline wires [were] of English pattern, landing wheels of Palmer type, bomb-releases . . . of their own design, and the duralumin machine-gun rings . . . of French pattern. Aircraft fuel pumps were the French A.M. type and mobile starters were the Hucks types."

At the request of the State Department and the Buckeye Pattern Works of Dayton, Ohio, the Secretary of War granted "release of Records of Tests made of certain aluminum exhaust stacks at the Aviation Depot at Wright Field, Dayton, Ohio, for benefit of the Russian Soviet Government." No military objections were made to the production of Wright aeronautical engines in Russia, and to an application by Sperry Gyroscope to sell bombsights. Nor was objection made to export of Type D-1 an D-2 oil bypass relief valves in 1935 by the Fulton Syphon Company of Knoxville. The Stupino plant also manufactured U.S. Hamilton 2-blade and 3-blade variable-pitch propellers for military aircraft.

The United Engineering and Foundry Company contracts of January 1938 exemplify the advanced nature of the aircraft materials technology supplied by Western firms to the Soviet Union. Indeed, some of these projects strained the research and development abilities of the most advanced Western firms and were far beyond the capability of the Soviet Union at that time. The contracts do suggest, however, that the Soviet Union has had remarkable ability to recognize advance military aircraft technology and enlist front-rank foreign firms in the acquisition process. The January 1938 United Engineering agreement involved the sale of $3 million worth of equipment and technical assistance for aluminum mills at Zaporozhe. These were 66-inch (1,680 mm.) hot and cold mills complete with auxiliary equipment — the most modern aluminum mills in the world. Jenkins, the United Engineering chief engineer in the USSR, said of the Zaporozhe mill that "not even the Aluminum Company of America has machinery as modern as it is." Both mills were "completely powered and controlled by General Electric apparatus."

The Stupino mill (Plant No. 150) near Moscow, by far the most important Soviet aluminum-development project, was also the subject of an agreement in May 1939 between Mashinoimport and United Engineering and Foundry for installation of hot and cold rolling mills. These were mills of extraordinary size to produce aluminum sheet for aircraft manufacture.

The Stupino installation comprised two sections: a hot mill and a cold mill. The hot mill had two units. One was a 2-high 66-inch hot rolling mill for rolling cast duralumin, including aircraft specification Type 17-S and 24-S ingots. The 66-inch mill came into regular operation about February 1, 1940 and the 112-inch mill a few weeks later. The cold mill contained two mills of similar size for cold working sheets produced in-the hot mill. The 66-inch cold mill started about March 1940 and the 112-inch cold mill late in 1940. All finishing equipment was supplied and placed in operation by United Engineering for the Soviets. The complete contract was worth about $3.5 or $4 million to United Engineering. For this sum the Soviets acquired an installation capable of rolling 2,000-foot aluminum sheets for aircraft. United Engineering said of it, "Nothing of such a size has ever been produced before."

During World War II the Soviets produced 115,596 aircraft from these materials and items of equipment while Lend-Lease delivered to the USSR only an additional 14,018. However, the Russian-produced aircraft were mainly obsolete prewar types and many were one-engine wood and canvas models with inferior engines. The full utilization of U.S. equipment came after World War II. Domestic production was assisted by a high degree of production specialization in a few plants with almost all foreign equipment. The only Soviet dive bomber, the Stormovik (IL-2), was in production at three plants; each plant produced about the same number of IL-2s, but not other aircraft. Fighter production was concentrated on the YAK-3, the YAK-2 and YAK-6 being advanced trainer versions. The YAK was produced in six widely scattered plants producing only YAK aircraft at a rate of between 65 and 400 per month.

Two-engined bomber production included the PK-2 (based on the French Potez) at two plants, and the IL-4 at three plants, of which only Komsomolsk (which Henry Wallace said was like the Boeing Seattle plant because it had all U.S. equipment) produced other aircraft. The LI-2 (Douglas DC-3) transport was produced only at Tashkent, and the PO-2 (or De Havilland Tiger Moth) was produced only at Kasan. Thus Soviet aircraft production was concentrated on a few types, each for a single flying function. Lend-Lease provided large quantities of advanced equipment for the development of the Soviet aircraft industry. Henry Wallace, after his visit to the important Komsomolsk aircraft plant, commented as follows:

> The aircraft factory [in Komsomolsk] where Stormovik bombers were built owed both its existence and its production to the United States. All the machine tools and all the aluminum came from America . . . It looks like the old Boeing plant at Seattle.

Foreign Designs for Soviet Aircraft Engines

By acquiring rights to manufacture foreign engines under license and with Western technical assistance, the Soviets were able to acquire a sizable engine-producing capacity for high-quality engines at low cost. For example, in the 1930s, Plants No. 24 and 25 were built in Moscow. Plant No. 24 made Wright Cyclone engines under license and Plant No. 25 made parts for Wright engines. Table 6-2 summarizes Soviet production of aircraft engines in 1940. All Soviet engines were foreign models produced under license.

Before this production program was established, prototypes of every Western aircraft engine were purchased (or stolen). These acquisitions were minutely examined and copied, or composition "Soviet" designs were built incorporating the best features of several foreign engines. A report by Bruce Leighton of Curtiss-Wright describes one of these early Soviet models at the Engine Research Institute in 1931:

> They've taken Packard, Conqueror, Rolls-Royce, Kestral, Hispano-Suiza, Fiat, Isetta-Franchini — tested them all, analyzed them down to the minutest details, including microphotographs of piston rings, flow lines in crank shafts, etc., taken good features of all, added some ideas of their own (particularly regards valve cooling) and built-up [sic] an engine which we're going to hear more of or I miss my guess.

In 1944, in the entire world, there were about 130 basic types and 275 variations of aircraft engines, either in production or recently in production. Each of the three Soviet engine types was an adaptation of a foreign engine built under a licensing agreement. The M-63 liquid-cooled 9-cylinder radial model was developed from the 1936 M-25, in turn developed from the Wright Cyclone, and used in the Soviet Consolidated Catalina patrol plane. The M-88 was a 14-cylinder air-cooled radial engine based on the French Gnome-Rhone 14-N, used in DB bombers, SU dive bombers, and PS transport planes. The third engine type was the M-105, a 12-cylinder liquid-cooled V-type of 1,100 horsepower based on the French Hispano-Suiza 12-Y engine, and used in the PE dive bombers, YAK fighters, and L-760 transport planes.

TABLE 6-2

Soviet Aircraft-Engine Production (1940)

Plant	Model No. of Engines Manufactured	Western Licenser of Engine Manufactured	Monthly Production
Aircraft motor works No. 29 Baranov	M-85, followed by M-87B and M-88	Gnome et Rhone (France)	130
Aircraft motor works No. 24 Frunze and No. 25	M-25, then M-63 and M-64	Curtiss-Wright (U.S.A.)	250
Aircraft motor works No. 26 Aviatroi Pavlov	M-100, M-103, then M-105P and M-105R	Hispano-Suiza (France)	270
Aircraft motor works No. 10, Tula	M-17, then M-38	BMW (Germany)	Not known

Source: German OKW files.

The Wright Cyclone Engine in the Soviet Union

In 1931 the Curtiss-Wright liquid-cooled engine was the only liquid-cooled American engine still in production. The U.S. Army initially supported development, but dissatisfied with the basic design, cut off funds in 1932. Development support for two other liquid-cooled engines, including the Curtiss-Wright H-2120, was continued by the U.S. Navy. Testing and development continued from 1933 to 1936, when the Navy withdrew support and reverted to air-cooled engines. The second U.S. Navy-supported Curtiss-Wright project was a 12-cylinder V-engine known as the V-1800. This was intended to replace the Curtiss-Wright Conqueror, and successfully completed its testing in 1934. Shortly after this test was completed, the Navy was forced by lack of funds to abandon most of its high-speed program and stopped support of the V-1800. The V-1800 engine was then licensed to the Soviet Union, which funded further research work to raise the engine rating to 900 horsepower from the U.S. Navy's test rating of 800 horsepower. This work was done at Aircraft Engine Plant No. 24 (Frunze) in Moscow, with parts manufactured at Plant No. 25. By 1938 these plants were producing about 250 Wright Cyclones (the Soviet M-25) per month. A plant for manufacturing Cyclone engines was also built at Perm — it was twice the size of the Wright plant in the United States.

The Soviet M-26 engine was based on the Pratt & Whitney Hornet. In July 1939 the corporation licensed the Soviet Union for production of the Pratt & Whitney Twin Wasp 1830 and the Twin Hornet 2180 aircraft engines.

The Gnome rotary, manufactured by the Societe des Moteurs Gnome et Rhone, was one of the finest early aircraft engines. After World War I the Gnome Company purchased the license of the British Bristol Jupiter-II and during the decade of the 1920s the Gnome-Rhone engineering department was dominated by English engineers from the Bristol Aeroplane Company. After producing the Bristol Jupiter engine for some years, the Gnome Company came up with an improved engine of its own design, using American lined cylinders. This crossfertilization of ideas led to the exceptional Gnome rotary engines of the 1930s, which were then adopted by the Soviets.

The Gnome-Rhone 114 was built at the Kharkov engine building plant (Plant No. 29) and the French engine became Soviet models M-85, M-87B, and M-88. About 1,500 M-88s a year were produced by 1940.

Similarly, the French Hispano-Suiza engine was produced in Moscow at an enormous plant twice the size of either the Pratt & Whitney or the Wright factories in the United States, themselves gigantic. This French Hispano-Suiza engine became the Soviet M-105 engine.

Western Contribution to the Postwar Soviet Air Force

In 1945 and 1946 the Russian aircraft industry concentrated on mastering the achievements of the German aircraft industry as it had been developed from 1941 to 1943. The years immediately after 1946 witnessed a remarkable expansion in the Soviet industry, based on this and on additional British technical assistance. Technical assistance from the West flowed in from the United Kingdom, particularly through transfer of the Rolls-Royce Nene, Derwent, and Tay jet engine technologies, and from Germany via the transfer of about two-thirds of the enormous German wartime aircraft industry to the Soviet Union.

The postwar aviation and space industries in the USSR have their roots in German World War II aircraft and rocket developments. In 1945 the Germans had a large and relatively undamaged aircraft and rocket manufacturing industry that had been dispersed under threat of Allied bombing toward the eastern regions of Germany — the area later occupied by the Soviets, or transferred to the Soviets on July 1, 1945. Over two-thirds of this productive capacty fell intact into Soviet hands and was removed to the USSR.

The major design units of the German aircraft industry, including most of the Junkers, Siebel, Heinkel, and Messerschmidt plants, were transported to Podberezhye, about 90 miles north of Moscow. Professor Walter Baade of Junkers continued development of the Ju-287K (as the EF-125) after moving to Podberezhye, and followed this with the T-140 and T-150 bombers. These were jets capable of carrying an atomic bomb and, according to one report, they could outperform the U.S. B-47. There were eleven major Junkers aircraft plants in the Soviet Zone and six of these are known to have been completely removed to the USSR, among them the main Otto Mader works, two miles east of Dessau (where Professor Baade had been located), and the Aschersleben, Bernburg, Leopoldshall, and Schonebeck plants. Aschersleben was a fuselage-building plant in process of changing over to the production of the jet He-162; its instrument storeroom, "virtually intact," was placed under military guard by the U.S. Army until the Soviets were able to take it over for removal to the Soviet Union. Benburg was also intact.

In 1944, the outstanding German rocket-plane designer Sanger was working on the Sanger-Bredt project to develop a long-range rocket aircraft. Former Russian General G. A. Tokaev recalls that in 1947 he was summoned to a Moscow conference at which Stalin said, "Von Braun, Lippish, Sanger, Tank and all kinds of other experts are working for the Allies, we must concentrate very seriously on German specialists."

Voznesensky then completed a draft decree, and read it aloud to the conference:

> The Council of Ministers of the U.S.S.R. decrees that a Government Commission shall be formed for the purpose of directing and co-ordinating scientific research into aviation problems, with special relation to piloted rocket planes and the Sanger Project. The Commission shall be composed of the following:
>
>> Colonel General Comrade Serov (President)
>> Engineer Lieutenant Colonel Comrade Tokavev (Deputy President)
>> Academician Comrade Keldysh (Member)
>> Professor Comrade Kishkin (Member)
>
> The Commission shall leave immediately for Germany, to undertake its preliminary work. A full report of its activities, and of the results it has attained, must be rendered to the Council of Ministers by August 1st.
>
> Marshall of the Soviet Union Comrade Sokolovsky is hereby directed to give the Commission every assistance.
>
> Moscow, the Kremlin, April 17, 1947[25]

"A thorough examination of the Sanger Project was invaluable," said Tokaev, because "of the experience such research would give our scientists in solving related problems and preparing a base for future ac-

[25]G. Tokaev, *Stalin Means War* (London: Weindenfeld & Nicolson, 1951), p. 158.

tivities. In other words, by mastering Sanger's theories our experts would be able to begin where he had left off.

Despite these high-level efforts, Professor Sanger was never captured by the Soviets, although the transfer involved almost all other German projects and technologies under development in 1945.

A troublesome gap in 1945 Soviet technology was modern fighter aircraft. Dr. Siegfried Gunther and Professor Benz, both developers of German fighter aircraft, were removed to the USSR. Gunther had been chief designer for Heinkel and a designer of jet fighters since the late 1930s, while Benz designed the German He-162 Volksjager jet fighter.

Among the Soviet acquisitions in Saxony was the Siebel works at Halle, where the experimental rocket-powered research aircraft DFS-346 (comparable to the U.S. Bell X-1 and X-2 and the Douglas X-3) was in final assembly. This work was continued at Halle on behalf of the Russians until Octoer 1948, when it was moved to the OKB-2 combine at Podberezhye with technicians from the Junkers, Heinkel, and Siebel plants. Flight testing of versions built in the USSR was begun in early 1948, using a Lend-Lease North American Mitchell B-25 bomber and later a Boeing B-29 Superfortress as mother aircraft. The first test pilots were Germans, later replaced by Russian pilots.

The aircraft-manufacturing facilities removed from Germany contained some unique equipment. Two German Wotan presses of 15,000 tons were taken and at least four copies were made in the Soviet Union and other units developed from these presses. Aircraft-equipment-manufacturing plants included the former Nitsche plant at Leipzig, used in the USSR to manufacture curve potentiometers, and the Karl Zeiss plant, for position-finders, wind-tunnel parts, and various precision instruments. It is estimated that in 1954 this segment of the wartime German aircraft industry supplied about 75 percent of Soviet radar equipment and precision instruments.

The Boeing B-29 Four-Engined Bomber becomes the Tu-4 and the Tu-70

During World War II the United States was unwilling to send four-engined heavy bombers to the Soviet Union under Lend-Lease. Although in April 1944 General John R. Deane recommended U.S. approval of Russian requests for heavy bombers, the War Department refused on the grounds that the Soviets could not train a bombing force prior to the spring of 1945 and that certain special equipment for such bombers was in short supply. The official State Department Lend-Lease report on war aid lists Russian acquistion of only one four-engined bomber (a B-24 that force-landed in Siberia), although the Soviets were in fact able to acquire four B-29s by retaining force-landed bombers in the Far East.

The Soviets then started work on the Tu-4 four-engined bomber and the Tu-70 civilian-transport version. In 1946 Amtorg attempted to purchase from the Boeing Aircraft Company a quantity of B-29 tires, wheels, and brake assemblies. In 1947 the Soviets produced the Tupolev Tu-70, which was immediately identified as a direct copy of the Boeing B-29. The similarity has been described in *Boeing Magazine:*

> The famed Boeing 117 airfoil on the Tu-70 is an exact replica of the Boeing B-29 wing. Along with the wing are the superfortress nacelles: outline, cooling air intake, auxiliary air scoop, cowl flaps and inboard and outboard fairings. The cabin cooling air inlet in the wing leading edge between the body and the inboard nacelle is the same. The trailing edge extension on the flap between the inboard nacelle and the side of the fuselage are also identical, according to the evidence provided by the photographs.

On the landing gear, Boeing commnts:

> The Tupolev Tu-70 uses the Twenty-nine's main landing-gear structure as well as its fairings and doors. The nose gear also appears to be that of the Superfortress, with the upper trunnion located closer to the body contour of the Tu-70 than on the Boeing bomber.

The tail surfaces of the Russian transport also came directly from the Boeing engineering department. On comparison it is apparent that the vertical tail and the dorsal outline as well as the leading edge of the rudder are the same on the two planes. The rudder of the Tu-70 appears to end at what would be the top of the tail-gunner's doghouse on the Superfortress. The shape of the stabilizer and the elevator is the same on the two planes, and the Tu-70 also uses the inverted camber of the B-29's tail.

The propellers of the Tupolev Tu-70 were original B-29 props, less cuffs. The hubs are characteristic of the Hamilton-Standard design. Boeing engineers also report that the drift-meter installation of the Russian transport looked like that of the superfortress, as did the pitot head type and location match-up.

The Soviets did design a new fuselage, higher on the wing of the Tu-70 than the fuselage of the B-29, larger in diameter, and a little longer (119 feet as compared to 99 feet). The Tu-70 transport retains the bomber nose, including the bombardier's plate-glass window.

How did the Soviets advance from an inability to produce modern bombers in 1940 to an ability to produce a workmanlike design requiring an extensive period of research and flight testing in 1947? Even if the finest designs were available, jigs and dies to put the plane into quantity production were also required. The 18-cylinder Wright engines for the B-29 had been extremely difficult to manufacture — even in the United States. Further, the Soviet's only experience in the production of four-engined bombers was the very unsuccessful Tupolev PE-8. We also know that in 1940 the Soviets had enormous difficulties putting the DC-3 twin-engined transport plane into production and repeatedly came back to the Douglas Aircraft Company for aluminum sections, parts, and technical advice.

Obviously, the record of a great deal of our assistance to the Soviets still lies buried in the U.S. government files. One area still worthy of research in the 1980s is the so-called "special programs" under Lend-Lease — unpublicized and still classified.

The First Soviet Jets

Aircraft Plant No. 1 at Kuibyshev, built by Lend-Lease during World War II, absorbed the equipment from the Junkers facility at Bernburg, which was transferred from Germany along with Junkers engineers, designers, foremen, and test pilots. The function of the plant was to utilize the emerging German jet technology to build the first Soviet jet fighters and bombers. The Soviet designers Tupolev and Gurevich began by visiting German aircraft factories to examine prototypes and production methods. The Junkers Company organized an exhibition of secret German aircraft projects and arranged for tours of inspection. Equipment was then removed under the program of OKSs (Osoboye Konstruktorskoye Byuro); OKB No. 1 was at the Junkers plant in Dessau.

The bulk of the German engineers and scientists were moved to Russia by train on the night of October 22-23, 1946, in what was probably the largest mass movement of scientific brains in the history of the civilized (or the uncivilized) world. These engineers and scientists were divided into small groups of about fifteen persons, with about thirty Russian engineers attached to each German nucleus for study and work. Each project was handled by stages — the draft stage, the technical project stage, and finally the presentation stage. Whenever a project was almost complete it was canceled by the Soviets and the related drawings, papers, biographies, and technical material were turned over by the Germans. Duplicate work was undertaken by separate all-Russian groups some distance from the location of the original German pilot-groups. In addition, German groups were put in competition with each other.

Most German designers and engineers in the aeroengine industry were sent to Kuibyshev Plant No. 1. They came mainly from the Junkers and BMW plants, no less than 800 engineers and technicians from these two companies alone. Among the members of the BMW contingent was Kurt Schell, former head of the BMW rocket laboratory, and engineers Winter, Kaul, Schenk, Tietze, Weiner, and Muller. The Junkers group led by Walter Baade was the most important. Dr. Baade, formerly chief engineer of Junkers with ten years experience in American aeronautical plants, was fully familiar with American methods of aircraft construction. With Dr. Baade was a group of engineers including Feundel, Haseloff, Wocke, Elli, Lilo, Rental, Hoch, Beer, Antoni, Reuss, Heising, and Hartmann.

The Junkers engine team in the Soviet Union was headed by Dr. Scheibe, who designed the Junkers P-1 turbine; he was assisted by engine designers Gerlach and Pohl, who at Dessau had been in charge of the engine testing department. Also in the Scheibe group were Steudel and Boettger and a large number of personnel from the Junkers turbojet department, including engineers, foremen, and skilled workers. Another prominent designer, Ernst Heinkel, worked in the Soviet Union at the Kalinin Experimental Station.

The Junkers plant itself was rebuilt at Kuibyshev, "almost exactly" as it had been in Leipzig.

Development of the First Soviet Jet Engine

At first German engineers were used to develop jet engines for the Soviets after World War II. First came reproductions of the Junkers-004 and the BMW-003 jet engines, which had been removed to the Soviet Union with their associated production equipment. The 004 became the Soviet RD-10, and the BMW-003 was produced as the Soviet RD-20 on a stop-gap basis until more advanced designs came along from British sources.

The first project given to the German design groups was a Soviet specification for a 3,000-horsepower jet engine, a development of the Junkers-012 turbojet, which had been in the design stage in Germany at the end of World War II. By 1947 the Junkers-012 had been developed as a 12-burner assembly, but operating inefficiencies halted development of this engine in 1948. The next project specification given to the German designers was for a 6,000-horsepower turboprop that could attain a speed of 560 miles per hour at sea level. This engine was developed from the Junkers-022 turboprop engine, with the same general design and characteristics as the 012.

By 1949 the Brandner design teams had essentially met the Soviets' specification and immediately set to work on yet another project — a power plant with 12,000 horsepower in contrast to the 6,000 horsepower developed by the Junkers-022. Finally, the Type-K turboprop was developed by the Junkers-BMW team as a 14-stage compressor and 5-stage turbine engine, a logical evolution from the German engines under development during the latter stages of World War II. Type-K engines produced by the mid-1950s power the operational Soviet four-engined bomber (Tu-20 Bear) with four MK-12M turboprop engines of 12,000 horsepower capacity, and the civilian version, Tu-114 (The Rossiya).

The AM series (after Mikulin) developed from the work of a Junkers-BMW team in the USSR under engineer Brandner. The end result of this design, the AM-3, was seen in 1958 by an American engineer, whose comment was, "The engine is not an outstanding power plant, being of simple design of very large diameter and developing about 15,000 pounds thrust with 8 compression stages."

The AM series of turbojets is currently used in the Tu-104 Camel civilian version of the Tu-16 Badger bomber.

MiG Fighters with Rolls-Royce Turbojets

In 1946 the Soviets bought fifty-five Rolls-Royce centrifugal compressor type turbojets — twenty-five Nenes and thirty Derwents. These Rolls-Royce engines, the most advanced in the world for the time, were well suited to Soviet production methods and introduced the Soviets to the use of a centrifugal turbojet. Up to 1947 Russian jets were all of the axial-flow type based on German designs. These Rolls-Royce turbines proved to be the best possible equipment for the MiG-15, which was designed by Siegfried Gunther and put into serial production under the name of the Soviet designers Mikoyan and Gurevich. Gunther was brought to Moscow and appointed chief designer in the construction office in Podberezhye.

Two versions of the Rolls-Royce engines were produced at Engine Plant No. 45 near Moscow from 1948 to the late 1950s. The plant was toured in 1956 by U.S. Air Force General Nathan Twining, who noted that it contained machine tools from the United States and Germany, and had 3,000 workers engaged in producing the Rolls-Royce Nene.

In 1951 the American counterpart to this Rolls-Royce engine was the Pratt & Whitney J-42 Turbo-Wasp, based on the Nene, but not then in quantity production. When the Korean War broke out in 1950, therefore, the Russians had thousands of improved Rolls-Royce Nene engines in service powering MiG-15s, whereas the U.S. Air Force had only a few hundred F-86A Sabres with comparable engines. Several engines from MiG-15s captured in Korea were evaluated by the United States Air Force. Reports were prepared by engineers of Pratt & Whitney Aircraft Division of United Aircraft Corporation, the Wright-Patterson Air Force Base, and Cornell Aeronautical Laboratory. We know from these analyses that by 1951 the Soviets had two versions of the original Rolls-Royce Nene in production quantities. The first version, the RD-45 that powered an early MiG-15, was a direct copy of the original Rolls-Royce Nene and delivered 5,000 pounds of thrust. The second version of the RD-45 delivered 6,000 pounds of static thrust at sea level and 6,750 pounds of thrust with water injection. The turbine blades in the Soviet RD-45 engines were made of a stainless steel alloy of the Nimonic-80 type, while the burner liner and swirl vanes were made of Nimonic-75. Parts of the Nene sold to Russia in 1948 were fabricated from Nimonic alloys — "Nimonic" being the registered trademark of Henry Wiggin and Company of Birmingham, England. Both Nimonic-75 and Nimonic-80 were developed by Mond Nickel about 1940, and the specifications had previously been published by the Ministry of Supply in the United Kingdom on the grounds that it was nonstrategic information.

The RD-45 (Nene) was produced in Moscow and also at Magadan from 1951 onwards, at Khabarovsk, at Ufa Plant No. 21, and at the Kiev Plant No. 43 from 1951 until sometime after 1958.

In 1967 the Soviet Strategic Air Force operated about 120 Tu-14 Bison bombers, 70 Tu-20 Bear bombers, and 1,000 Tu-16 Badger bombers. The Soviet Navy also operated these three types of aircraft.

From the information in Table 6-1 we can trace the operational jet engines of the 1960s from the Junkers and BMW prototypes taken from Germany at the end of World War II or from those sold by the Rolls-Royce Company as "peaceful trade" in 1946. Both groups of prototypes were developed by German engineers transferred to Russia as forced labor, wtih equipment and instruments imported as "peaceful trade." When the K-series and the AM-series turbojets were well along the development road the Germans were returned home. The Soviets have had no difficulty since in making design improvements to the original German and British concepts and technologies. These are the engines that power operational Soviet military aircraft.

Even today in the 1980s Soviet military and commercial aircraft are heavily dependent on Western technology — primarily American.

It has been observed that current aircraft designs from Soviet military aircraft design bureaus are to a significant degree copies of Western military aircraft. Apparently Soviet military designers order documents and plans from U.S. Government sources and we 'obligingly make deliveries within a month or so directly to the Soviet Union. For example, the Soviet military cargo plane is a direct copy of the U.S. C-5A giant transport. Soviet acquisition of American aircraft also suggests they are just as interested in aircraft construction techniques as in design data for particular aircraft.

This Soviet design interest has been identified especially in military transport and wide body jets and they have probably managed to accelerate their development programs significantly. The Soviet IL-86 is a copy of the Boeing 747. The IL-76 is a copy of the C-141. While neither is a precise copy, both designs are substantially the same.

Even the NASA Space Shuttle has been copied. In 1984 U.S. intelligence sources reported that the Soviet Union is building a "carbon copy" of the Space Shuttle. Retired Lt. Colonel Thomas Krebs, former chief of the DIA space systems branch, reported: "We've seen the Soviet orbiter and it's identical to ours."

The only difference between the shuttles is reported as an additional set of engines below the fuel tank, thus an increased payload capacity. The Soviets were able to purchase a complete set of Space Shuttle plans. These were unclassified and made available by NASA to any interested party. The Soviets were an obviously interested party, although it is beyond comprehension why the NASA people would release a technology with obvious military implications. The NASA excuse is that the plans were released to improve coordination with commercial suppliers of equipment. The Soviets set up a dummy company to make the purchase (*Washington Post*, June 9, 1984).

Table 6-1

Western Origin of Some Soviet Military Aircraft

Aircraft Model	Date in Service	Engine Utilized	Origin of Engine
MiG-9 fighter	1946-47	RD-20	BMW 003
MiG-15 fighter	1947-1960s	RD-45	Rolls-Royce Nene
MiG-17 fighter	1954 to date	VK-2JA	Rolls-Royce Nene
MiG-19 fighter	1955 to date	VK-5 or (M-205)	Rolls-Royce Tay & Derwent
Tu-70 (Boeing B-29)	1950	4 piston-type engines	Wright 18-cylinder
Tu-16 Badger bomber	1954 to date	AM-3M turbojets	Junkers-BMW team
Tu-104 airliner version of Badger bomber	1957 to date	AM-3M turbojets	Junkers-BMW team
Tu-20 Bear bomber	1955 to date	NK-12M turbo-props	Junkers-BMW team
Tu-114 airliner version of Bear BOMBER	1957 to date	NK-12M turbo-props	Junkers-BMW team
Cargo transport	1960	US C5A	
IL-86	1970	US Boeing 747	
IL-76	1970	US C-141	
Space Shuttle	1980	US NASA Shuttle	

The Supersonic Tu-144 (Alias "Konkordskiy")

British and French aeronautical engineers have their own name for the Russian Tu-144 supersonic plane. They call it the "konkordskiy." A comparative glance at the configurations of the Anglo-French Concorde and the Russian Tu-144 will — even without supporting evidence — readily explain the nickname.

The configurations of the Russian supersonic Tu-144 and the Anglo-French supersonic Concorde are strikingly similar. Given the history of Soviet technical dependence on the West, we can pose the question: Did the Soviets use the design of the Anglo-French Concorde for the Russian Tu-144?

Design work for Concorde began a decade before the British and French signed the Concorde agreement in 1962. Wind-tunnel testing, which yielded the data for the shape of the plane, began in the early 1950s. The Soviets had many other pressing problems in the early 1950s that were more important than research on a supersonic delta plane. However, the Tu-144 has a design concept very close to that of the Concorde. Both have modified double-delta wings, fixed geometry and low-aspect ratio for minimum drag. Fins and rudders are similar; neither aircraft has tailplanes. The major external differences are relatively slight variations in landing gear and engine position. In other words, superficially the Tu-144 is quite unlike anything the Soviets have designed previously; it is a significant jump in the technological horizon (but not as much as the aborted titanium U.S. supersonic plane) and should have required many years of testing and design work.

Dr. William Strang, technical director of British Aircraft Corporation's commercial aircraft division, has stated, "I think it likely that they did have some knowledge of the work we were doing which led to the general shape definition" (*London Times*, Sept. 27, 1971).

In September 1971 the British government expelled 105 Russian "diplomats" from England on charges of spying, and specifically military and industrial spying. According to the *London Times*, this espionage included "information on electronics, transformers, semi-conductors, computer circuitry, and technical details of the Concorde and Olympus 593 engine" (Sept. 25, 1971).

Finally, Doyle, a reformed member of the British Communist Party, confessed to accepting L5,000 from the Soviets for information on Concorde, "including manuals, sketches and small pieces of equipment." Security was so lax at the plant that Doyle and his Soviet friends once considered smuggling out a 16-foot missile disguised as a telegraph pole. This was no real problem, as Doyle had keys to all secret departments and security was nonexistent, but he balked at having to answer to his chief for a missing missile. Concorde was one thing, a missile was something else.

British and French engineers may have some justification for renaming Tu-144 the "Konkordskiy."

The SS-20 Guidance System Has U.S. Technology

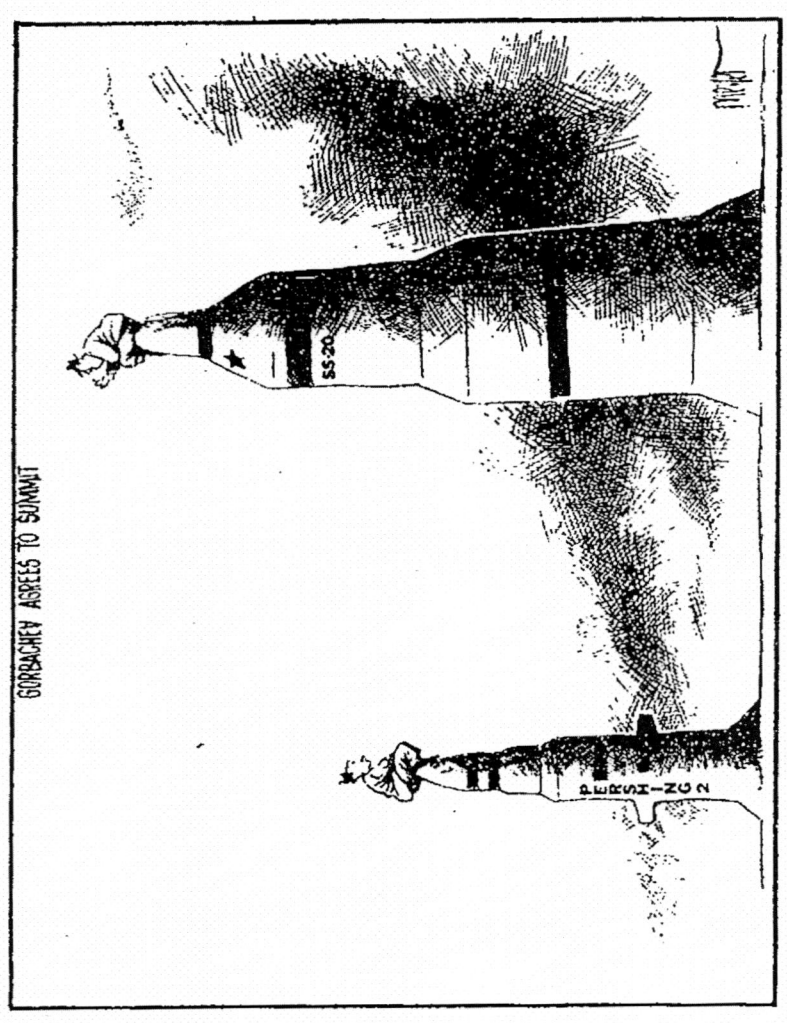

CHAPTER VII

The Deaf Mutes and the Soviet Missile Threat

"As for businessmen, I could persuade a capitalist on Friday to bankroll a revolution on Saturday that will bring him a profit on Sunday even though he will be executed on Monday."

Saul Alinsky, Chicago professional activist

The United States and the Western world today face a truly awesome threat from Soviet missiles. This threat would not exist if President Richard Nixon and National Security Adviser Henry Kissinger had heeded warnings in 1970 from its own Department of Defense and outside experts that the Soviets were lagging in missile production technology and required specific technologies from the West to MIRV their fourth generation ICBMs.

MIRV capability is the ability to deploy a number of warheads from the same missile, thus vastly increasing throw weight. Soviet third generation missiles did not have this capability. As stated by a Department of Defense report: ". . . it was not until the fourth generation that the technology became available to the Soviets allowing greater throw weight and greatly improved accuracy so that high yield MIRVs could be carried by operational missiles"

The phrase "became available" is a subtle way for DOD to state what has been concealed from the public: that the U.S. made the technology available (as we shall show below). The fourth generation ICBMs are the SS-17, the SS-18 and the SS-19, which today have the capability to destroy most of our 1,000 U.S. Minuteman missiles now operational with only a portion of their warheads.

American Accelerometers for Soviet Missiles

Let's go back to the start of our help for the Soviet missile program.

Accelerometers are small but vital instruments used in missiles and aircraft to measure gravitational pull. In 1965-68 the Soviets displayed an extracurricular interest in American accelerometers, and a Soviet United Nations diplomat was forced to hurriedly leave the United States before being picked up for espionage involving acquisitions of U.S. accelerometers.

Testimony of Leonard I. Epstein, vice president of Trans-American Machinery and Equipment Corporation of New Jersey, to the House Un-American Activities Committee detailed Soviet interest in this American technology. Mr. Epstein related to the committee how he met Vadim Isakov, a Russian employee of UNICEF (United Nations International Children's Emergency Fund) on July 15, 1965, and how Isakov

Soviet Fourth Generation ICBM's

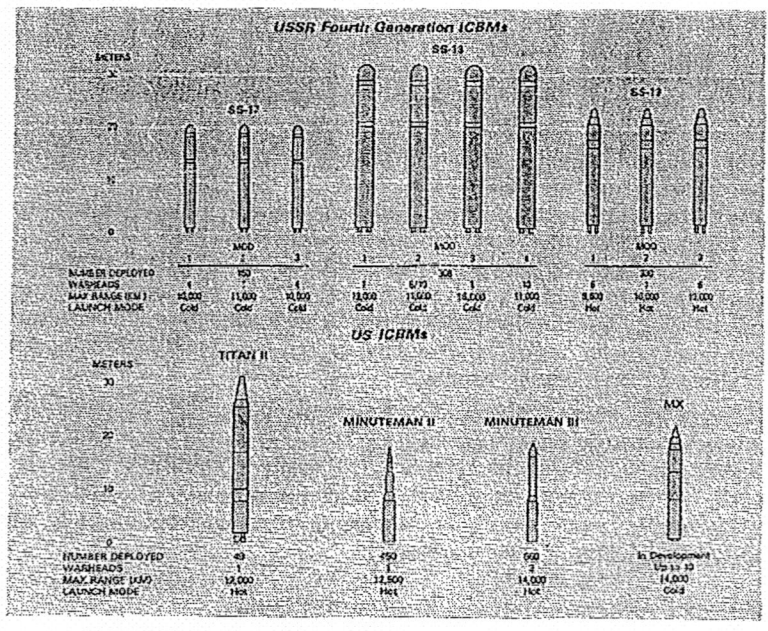

Source: Department of Defense, *Soviet Military Power* 1983 Washington, D.C.

later visited Epstein's plant in New Jersey with a list of four items for purchase, including "an accelerometer made by American Bosch Arma Corporation or similar company. The accelerometer is an intricate device which measures the pull of gravity on any vehicle such as a missile or space-orbiting device. The device costs about $6,000." Mr. Epstein, under instructions from the FBI, met several times with Isakov to "find out what he wanted."

In October 1965, "Isakov began to push for delivery on the accelerometer. [Epstein] surmised that the urgency had something to do with the fact that the Soviets had smashed three vehicles onto the surface of the moon." Although Epstein was able to stall for "quite some time" on various grounds, Isakov later became "quite anxious to obtain an accelerometer." When Epstein pleaded export problems, Isakov suggested he would use the Soviet diplomatic pouch.

Eighteen months later, in August 1967, another Russian, intensively interested in accelerometers, turned up in the United States, this time under the auspices of the State Department Academic Exchange Program. From August 1967 to June 1968, Anatoliy K. Kochev of the Kalinin Polytechnical Institute of Leningrad was at Catholic University in the United States working on "construction methods of equipment to measure small accelerations and displacement," that is, the manufacture of accelerometers.

Is there any connection between Isakov's unsuccessful espionage attempts to purchase accelerometers and Kochev's "academic" work on accelerometer manufacture in the United States, courtesy of the State Department? There are indeed obsolete accelerometers and sophisticated accelerometers. The Soviets know the difference. They know how to make the obsolete versions, but do not have the technical ability to make more sophisticated instruments. The trick is in the manufacturing process — that is, in knowing how to build into the instrument the sensitivity necessary to measure small gravitational pulls quickly and accurately. **It is the manufacturing technique that was important to the Soviets — much more important than a boatful of purchased accelerometers.**

Why did Kochev come to the United States in 1967? The State Department reports the title of his project as "construction methods of equipment to measure small accelerations." Ten months would be sufficient time for a competent engineer to determine the most modern methods in this field, and given the rather careless manner in which advanced accelerometers have found their way into used electronic equipment stores, it is unlikely that Kochev had major problems in adding to his knowledge of the state of the art.

Why did the State Department make an agreement in 1966 to allow a Soviet engineer into the United States to study the **manufacture** of accelerometers only a few months after another Soviet national had been foiled by the FBI in attempting to **purchase** an accelerometer? We have no answer for that.

American Ball Bearings for Missile Guidance System

In the late 1960s Soviet missiles were extremely inaccurate. According to Abraham Shifrin, a former Defense Ministry official, they could hardly find the United States, let alone a specific target. By the late 1970s their accuracy was so improved that Soviets could guarantee a high proportion of hits on a target as small as the White House.

The technological roadblock was mass production of miniaturised precision ball bearings for guidance systems.

In the early 1960s Soviets attempted to buy U.S. technology for mass production of miniaturised precision bearings. The technology was denied. However, in 1972 the necessary grinders were sold by Bryant Chucking Grinder Company and its products are today used in Soviet guided missile systems and gyroscopes. Specifically, the Soviets were then able to MIRV their missiles and increase their accuracy.

This is how the tragedy came about.

Ball bearings are an integral part of weapons systems, there is no substitute. The entire ball bearing production capability of the Soviet Union is of Western origin — utilizing equipment from the United States, Sweden, Germany, and Italy. This transfer has been fully documented elsewhere by this author (see Bibliography). All Soviet tanks and military vehicles run on bearings manufactured on Western equipment or copies of Western equipment. All Soviet missiles and related systems including guidance systems have bearings manufactured on Western equipment or Soviet duplicates of this equipment.

One firm in particular, the Bryant Chucking Grinder Company of Springfield, Vermont, has been an outstanding supplier of ball bearing processing equipment to the Soviets. In 1931 Bryant shipped 32.2 percent of its output to the USSR. In 1934, 55.3 percent of its output went to Russia. There were no futher shipments until 1938, when the Soviets again bought one-quarter of Bryant's annual output. Major shipments were also made under Lend-Lease. Soviet dependence on the West for ball bearings technology peaked after the years 1959-61, when the Soviets required a capability for mass production, rather than laboratory or batch production, of miniature precision ball bearings for weapons systems. The only company in the world that could supply the required machine for a key operation in processing the races for precision bearings (the Centalign-B) was the Bryant Chucking Grinder Company. The Soviet Union had no such mass production capability. Its miniature ball bearings in 1951 were either imported or made in small lots on Italian and other imported equipment.

In 1960 there were sixty-six Centalign-B machines in the United States. Twenty-five of these machines were operated by the Miniature Precision Bearing Company, Inc., the largest manufacturer of precision ball bearings, and 85 percent of Miniature Precision's output went to military applications. In 1960 the USSR entered an order with Bryant Chucking for forty-five similar machines. Bryant consulted the Department of Commerce. When the department indicated its willingness to grant a license, Bryant accepted the order.

The Commerce Department's argument for granting a license turned on the following points: (1) the process achieved by the Centalign was only a single process among several required for ball bearing production, (2) the machine could be bought elsewhere, and (3) the Russians were already able to make ball bearings.

The Department of Defense entered a strong objection to the export of the machines on the following grounds:

> In the specific case of the granting of the export license for high-frequency grinders manufactured by Bryant Chucking Grinder after receiving the request for DOD's opinion from the Department of Commerce, it was determined that all of the machines of this type currently available in the United States were being utilized for the production of bearings utilized in strategic components for military end items. It was also determined from information that was available to us that the Soviets did not produce a machine of this type or one that would be comparable in enabling the production of miniature ball bearings of the tolerances and precision required. A further consideration was whether machines of comparable capacity and size can be made available from Western Europe. In this connection, our investigation revealed that none was in production that would meet the specifications that had been established by the Russians for these machines. In the light of these considerations it was our opinion that the license should not be granted.

The Inter-Departmental Advisory Committee on Export Control, which includes members from the Commerce and State Departments as well as the CIA, overruled the Department of Defense opinion, and "a decision was made to approve the granting of the license." The Department of Defense made further protests, demanding proof that either the USSR or Western Europe was capable of producing such machines. No such proof was forthcoming.

The following is a summary of the objections of the Department of Defense representative:

> (a) I expressed dissatisfaction and suggested that the Department of Defense not concur in the initial request of the Department of Commerce.

(b) The official member of the Department of Defense in this connection concurred and, at a series of meetings of the Advisory Committee on Export Control, spoke against the proposal that an export license be granted.

(c) The Deputy Assistant Secretary of Defense Supply and Logistics, after reviewing some of the circumstances, requested that I do whatever was possible to stop the shipment of these machines.

(d) A letter was transmitted from the Office of the Secretary of Defense to the Secretary of Commerce, approximately November 1, 1960, saying it [sic] spoke to the Department of Defense and requesting a further review.

(e) At two meetings where the matter was reviewed, the Department of Defense maintained nonconcurrence in the shipment of the equipment.

As of this writing I am still convinced that it would be a tragic mistake to ship this equipment.

The reference to a "tragic mistake" refers to the known fact at that time that miniature precision ball bearings are essential for missiles. Granting the license would give the USSR a miniature ball bearing production capability equal to two-thirds that of the United States.

In 1961 a Senate subcommittee investigated the grant of this license to Bryant. Its final report stated:

The Senate Subcommittee on Internal Security has undertaken its investigation of this matter not in any desire to find scapegoats, but because **we felt that the larger issue involved in the Bryant case was, potentially, of life-or-death importance to America and the free world.** We are now convinced, for reasons that are set forth below, that the decision to grant the license was a grave error.[26]

The testimony of Horace Gilbert to the Senate summarizes the position on the Centalign machines;

Mr. Chairman, I am Horace D. Gilbert, of Keene, N.H., and I am president of Miniature Precision Bearings, Inc., and I would like to express my appreciation for having an opportunity to be here with you and come particularly at this time, when I know that everyone is so busy, and at such short notice. As the name implies, my company produces miniature ball bearings of precision quality, 85 percent of which are used in the national defense effort. All but 1 percent of our sales are within the United States, and most of these bearings are produced by machines manufactured by Bryant Chucking Grinder Co., of Springfield, Vt.

[26]U.S. Senate, *Proposed Export of Ball-Bearing Machines to U.S.S.R.* (Washington, 1961).

Our company owns about 25 of these machines out of the 66 which, I believe, presently exist in the United States. This machine was developed over a long period of years, and much of the know-how, Mr. Chairman, in the latest model, was contributed by our company.

Several months ago Russia ordered 45 of these machines from Bryant, and the Department of Commerce has granted an export license.

I was very much disturbed when I learned of this, and I and Mr. Patterson over there — whom I will further identify as one of the developers of this machine — we have attempted to demonstrate to the Department of Commerce the tragedy of these machines being sold to Russia.

Unfortunately, we have not met with success, and I would like to assure you, Mr. Chairman, that if these machines are sold, it means absolutely no commercial or financial difference to us as a company or to me as an individual.

I have no fear as far as Russia selling in our markets is concerned, and our company does not do any significant amount of bearing business in their markets. I am here because I think that this is folly which would undermine our defenses.

The Department of Commerce has attempted to justify its decision with four or five arguments, none of which, in our opinion, appears to be valid, and I would like to touch on these.

First, they say these machines could be purchased in Europe, and consequently, Bryant might as well benefit by their sale here.

I am thoroughly familiar with the machines which are in production in Europe. Part of my knowledge has been gained by three trips to Europe in the last 11 months, and I can assure you that no European manufacturer in fact does produce comparable machines with the accuracy of that which is used by Bryant. I would suggest that, if the Russians could buy this machine in any other market, they would indeed do so. In fact, an American competitor of Bryant, Heald Machine Co., of Worcester, has been attempting for 3 years to imitate and produce a comparable machine, and they have not been successful.

Secondly, the Department of Commerce has pointed out that 45 of these machines which have been ordered by the Russians are only a small part of the total number in existence. The number in existence is a matter of record, and, as of the end of September, there were only 66 machines in the United States, and now Russia has ordered 45 for their own needs, and I understand that not all 66 of these are in production. They are in experimental facilities, they would have almost the equivalent of the entire U.S. capacity for production.

Thirdly, the Department of Commerce has suggested that these machines require skilled operators who need substantial training; and I can assure you this is not true, sir. Even if it were, I am confident that the Russians have skilled technicians who, in a short time, would be able to master the operation of this machine, were it complicated, which it is not, and that is part of the magic of the machine, that it is not complicated. There is a certain amount of skill required to set up the machine, but under a contract with Bryant, I understand that the machine must be disassembled and reassembled in the presence of Russian inspectors who are not at their doorstep. Consequently, they will have whatever knowledge they need to put this machine into immediate operation.

The case for Bryant Chucking Grinder Company is expressed in the following portions of a letter sent to Senator Dodd on January 27, 1961 by N.A. Leyds, Bryant's vice president and general manager:

We appreciate the opportunity to make the following remarks concerning the testimony of December 21, 1960 and January 24, 1961, received by your Committee relative to the license granted to us by the Department of Commerce for the shipment of 45 of our Model "B" Centalign machines to Russia. . . .

There was no objection by any of the dissenters to the shipment of the J&L machines, and we certainly have no objection. But it has been readily admitted that these machines will quite probably be used in the production of miniature bearings. . .

We were not surprised at the objection by the Department of Defense as it is well known that their technical expert, who could not appear, is, and has been, against the shipment of most, if not all, machine tools to Russia. We do not question his sincerity nor wish at this time to discuss the validity of this person's opinion, but, whether this opinion described the policy of U.S. Government in this area is highly questionable. To our knowledge, the top technicians from the Defense Department have not seen our Model "B" machine. . .

We, along with other machine tool builders, are not restricted from producing any of the machines, not on the international control list, in our foreign subsidiaries. Our Company has subsidiaries producing machine tools in England and West Germany and presently regulations permit us to manufacture Model "B"s with spindle speeds up to 120,000 r.p.m. in these subsidiaries and ship them to Russia . . .

Our leadership in technology in this area is so slight that we must continually utilize our forces and talents at the maximum, to maintain the slightest gap which exists. We must be permitted to compete with our foreign competitors and maintain a healthy posture, or we must rapidly lose the race to maintain superior technology. A few other key machine tool builders with a similar problem can create a situation with far-reaching consequences to the nation's security. It is only when a company is strong that it can support the financial burden necessary to maintain research and development activities at their proper level.

In general we believe that in the matter of trade with the Soviet bloc, similar restrictions should apply to identical industries in each and all of the Free World countries. Our hands must not be tied.

The Senate subcommittee's conclusions were overwhelmingly in favor of denying the export license and raised major unanswered questions concerning the intentions of Leyds, the Bryant Chucking Grinder Company, and the Department of Commerce. These were the subcommittee's conclusions:

We believe that this testimony gives overwhelming support to the stand taken by the Department of Defense in this matter, and to the arguments presented by Miniature Precision Bearing in opposing shipment.

This testimony establishes conclusively (1) that the miniature bearings produced with the help of the Bryant machine are used primarily for defense purposes; (2) that the function performed by the Bryant machine is of critical importance; (3) that no comparable machines can at present be obtained from other sources; (4) that Soviet industry has not been able to master the problems involved in mass producing high precision miniature bearings; that the industry is in fact plagued by poor quality and obsolete equipment; that with its own resources, it would probably take a number of years to develop the capability; (5) that the possession of these machines would greatly accelerate Soviet mastery of the art of miniaturization. . . . we think it would be helpful if we briefly summarized some of the high points of this testimony and recapitulated some of the essential facts.

1. At least 85 percent of the bearings manufactured with the help of the Bryant machine are used by defense industries:

Subject machine is a key factor in the economical production of the highest precision for many important Department of Defense applications, such as the latest guidance systems, navigation, fire control, computer, synchro and servo mechanisms used for aircraft, ordnance, ships, missiles and other space vehicles (statement of Mr. J.R. Tomlinson, president, and Mr. B.L. Mims, vice president in charge of engineering, the Barden Corp., Danbury, Conn.).

2. **The function performed by the Bryant machine is of critical importance:**

The outer ball track grinding operation is one of the last and most vital of those performed on the bearing outer ring. It is the operation which, until the advent of this machine, could probably be called the bottleneck opposing the precision performance of miniature bearings. The necessary perfection of other operations has been achieved 5 to 20 years ago (statement by Mr. H.B. Van Dorn, vice president in charge of engineering, Fafnir Bearing Co., New Britain, Conn.).

3. **The Bryant machine is unique in its field:**

Secretary Mueller in his letter of January 18, 1961 to Senator Dodd, said that "substantially comparable" machines could be obtained from other sources. Mr. Bradley Fisk, Assistant Secretary of Commerce for International Affairs, in his testimony before the subcommittee on January 24 said that there are "five factories outside of Russia that could make similar machines." It was not clear from his statement whether the companies he named do, in fact, make such machines, or whether they are theoretically capable of making them. A careful check has revealed that none of the companies named by Mr. Fisk produce machines that can be considered equal or "substantially comparable" to the Bryant machine.

For the Soviets and Bryant Chucking Grinder Company the matter did not end in 1961.

In 1972, just before the presidential election, Nicholaas Leyds, general manager of the Bryant Chucking Grinder Company, announced a contract with the Soviets for 164 grinding machines. Anatoliy I. Kostousov, Minister of the Machine Tool Industry in the Soviet Union, then said they had waited twelve years for these machines, which included mostly the banned models: "We are using more and more instruments of all kinds and our needs for bearings for these instruments is very great. In all, we need to manufacture five times more bearings than 12 years ago."

Under President Nixon and National Security Adviser Henry Kissinger license for export of these 164 Centalign-B machines was approved.

Simultaneously came heavy pressure on this author to stop research work on our technological aid to the Soviet military system and to stop making public speeches on this aid.

By 1974 the Soviets had MIRVed their missiles and were in mass production. The results we well know and are reflected in the chart on page

On March 8, 1983, Secretary of Defense Weinberger not only made this massive Soviet increase public, but admitted something not admitted in the early 1970s: that the newly achieved accuracy was derived from our U.S. technology.

> We see from the strategic forces that the Soviets have dramatically increased their offensive strategic capabilities. The number, the explosive power and the accuracy of their ICBMs, an accuracy which, as we've said many times, has been largely derived from technology they have taken from us, this is far greater than they would need to simply deter the attack. The hardening of their silos, their provisions for reloading some of their larger ICBMs, a reload capability, refire capability which we do not have, and their enhanced strategic defenses, together with all of their writings and their exercises and the funds they've spent on civil defense, all of that suggests that they are developing the capability, and believe they are developing the capability which is equally important, of fighting a prolonged nuclear war. It is essential that, as we strive to maintain our deterrent.

CHAPTER VIII

The Soviets at Sea

"Within weeks many of you will be looking across just hundreds of feet of water at some of the most modern technology ever invented in America. Unfortunately, it is on Soviet ships."

> Secretary of the Navy John Lehman,
> May 25, 1983, to graduating class at
> Annapolis (reported in U.S. Naval
> Institute Proceedings, August 1983, pp. 73-4)

Only one Soviet battleship was built before World War II — the **Tretii International** ("Third International"), laid down on July 15, 1939 in the Leningrad yards. The guns, turrets, armor, and boilers for this 35,000 ton battleship were purchased in the United States and Germany. The ship was completed in the late 1940s. Other prewar Soviet battleships — the **Marat, Kommuna,** and **Oktyabrskyaya Revolutsia** — were reconditioned and refitted ex-tsarist vessels. Attempts to build three battleships of the Italian **Vittorio Veneto** class were abandoned.

Three aircraft carriers were under construction at the end of the 1930s. The **Stalin** (formerly the tsarist **Admiral Kornilov**), a 9,000-ton ship built in 1914, redesigned in 1929, and completed in 1939 as an aircraft carrier. Two other carriers of 12,000 tons each were built "on the basis of American blueprints" — the **Krasnoye Znamye** and the **Voroshilov,** laid down at Leningrad in 1939 and 1940.

World War II Soviet cruisers were refitted tsarist-era vessels, including the **Krasni Kavkaz** (formerly the **Admiral Lazarov,** built in 1916 at Kinoleav), the **Profintern** (formerly the **Svetlana,** built in 1915 at Reval [now Tallinn] and refitted in 1937), and the **Chevonagy Ukraina** (formerly the **Admiral Nakhimov,** built in 1915). The first Soviet attempt at cruiser construction was the **Kirov** class of 8,000 tons. Three ships were laid down in 1934-35 with Tosi engines manufactured in Italy and built to Italian plans at Putilovets (the **Kirov** and **Maxim Gorki**) and at Nikoleav (the **Kuibyshev**) under the technical direction of Sansaldo, an Italian firm.

There were three groups of Soviet destroyers before World War II. First, fourteen tsarist vessels — four in the **Petrovski** class (built in 1917-18), nine in the **Uritski** class (built in 1914-15), and one ex-**Novik** (built in 1911). Second, some new classes of destroyers were built under the Soviets to French and Italian designs. Between 1935 and 1939, fifteen destroyers of 2,900 tons each, based on French drawings, were built as the **Leningrad** class: six in the Leningrad yards, eight on the Black Sea, and one at Vladivostok. The first units, supervised by French engineers, were similar to French vessels.

The third category encompassed the **Stemitelnie** class, the largest Soviet destroyer class of the 1930s. Between 1936 and 1939, thirty-five of these 1,800-ton ships were built under Italian supervision, mainly in Leningrad and the Black Sea yards, utilizing an Italian Odero-Terni-Orlando design and British machinery. Their engines were Tosi (Italy) 50,000-shaft-horsepower geared turbines. In addition, the **Tashkent**, another Odero-Terni-Orlando design, was built in Italy — the only Soviet surface warship built abroad in the 1930s.

In January 1939 the American firm of Gibbs and Cox, naval architects, designed two destroyers and a 45,000-ton battleship for the Soviet Union in the United States.

From 1939 to 1941 the Soviets received German military assistance. The Nazis sent the partly finished cruiser Lutzow, laid down at Bremen in 1937, and in May 1941 "construction of the cruiser 'L' in Leningrad was proceeding according to plan." In the Leningrad yards German technicians took over construction and repair of Soviet ships. This cooperation lasted for eighteen months, from late 1939 until May 1941.

All told, in 1941 the Soviet fleet comprised 3 battleships, 8 cruisers, 85 destroyers and torpedo boats, 24 minelayers, 75 minesweepers, 300 motor torpedo boats and gunboats, and 250 submarines. Most were built in the West or to Western designs.

U.S. Lend-Lease added 491 ships to this total: 46 110-foot submarine chasers and 59 65-foot submarine chasers, 221 torpedo boats (24 of them from the United Kingdom), 77 minesweepers, 28 frigates, 52 small landing craft, 2 large landing craft from the United Kingdom, and 6 cargo barges. In addition to combat vessels, Lend-Lease transferred merchant ships and marine engines.

In terms of tonnage, Lend-Lease probably doubled the size of the Soviet Navy. Only a small number of these naval ships have been returned, although the Lend-Lease master agreement required the return of all vessels.

Since World War II, assistance to the Soviet naval construction program has taken two forms: export of shipbuilding equipment and shipyard cranes from European countries and the United States, and use of plans and designs obtained from the United States and NATO through espionage. For example, the sophisticated equipment of the U.S.S. **Pueblo**, transferred by the North Koreans to the USSR, was at least fifteen years ahead of anything the Soviets had in the late 1960s. In other words, the **Pueblo** capture took the Soviets in one leap from postwar German and Lend-Lease technical developments to the most modern of U.S. technology.

Current Soviet acquisitions of naval equipment are highly significant and evidence of failure by the West to maintain a realistic defense posture.

The Soviets have concentrated their acquisitions in areas related to aircraft carriers, deep sea diving capabilities, sensor systems for antisubmarine warfare and navigation, and ship maintenance facilities. In the maintenance area, two huge floating drydocks purchased from Japan, supposedly for civilian use, have been diverted to military use. Drydocks are critical for both routine and fast repair of ships damaged in warfare. In 1978, when the Soviets took possession of one of the drydocks, they diverted it to the Pacific Naval Fleet. The other was sent to the Northern Fleet in 1981.

These drydocks are so large that they can carry several naval ships. More importantly, they are the only drydock facilities in either of the two major Soviet fleet areas — Northern or Pacific — capable of servicing the new Kiev-class V/STOL aircraft carriers. Soviet advanced submarines carrying ballistic missiles, Soviet Kiev aircraft carriers, and Soviet destroyers were among the first ships repaired in these drydocks. The drydocks are so large that no Soviet shipyard is capable of accommodating their construction without major facility modifications, associate capital expenditures, and interruption of present weapons programs. Their importance will be even more pronounced when the Soviets construct the still larger carriers (for high-performance aircraft) projected for the 1990s. The Soviets have acquired Western aircraft carrier catapult equipment and documentation for this larger carrier; catapult technology, though relatively common in the West, is outside Soviet experience and capabilities.

In the 1980s, the USSR has contracted for or purchased foreign-built oceanographic survey ships equipped with some of the most modern Western-manufactured equipment. In place of U.S. equipment that was embargoed, other Western equipment has been installed on the ships. This modernization of the world's largest oceanographic fleet with Western technology will support the development of Soviet weapon systems programs and antisubmarine systems used against the West.

Ship and submarine construction requires sheet steel, steel plate, and steel sections. Armor-plate is produced by rolling high-alloy steel, which is then heat-treated to develop its ballistic properties. Multiple layers are used for armor protection. Therefore, assistance to the Soviet iron and steel industry — which is significant and continuing — is also assistance to Soviet ship-construction programs.

A U.S. government agency report has asserted that "any shipyard capable of building a merchant ship hull is equally capable of building a combatant ship of the same length." The report also states that merchant ships can be designed for conversion into naval ships, and that in any event the facilities required to build a steel merchant ship are exactly the same as the facilities required to build a steel warship. The main differences are the armament and the varying specifications for engines and other equipment. Almost 70 percent of the present Soviet merchant fleet has been built outside the Soviet Union. This has released Soviet shipyards and materials for Soviet naval construction. All diesel engines in Soviet ships use a technology originating outside the Soviet Union.

The Soviets provided 80 percent of the supplies for the North during the war in Vietnam. Most of these supplies were transported by merchant ship. The ocean-going capacity required to supply the North Vietnamese on this scale and so keep them in the war was dependent upon ships previously built outside the USSR. The same process can be identified in Central America and Africa. While the supply of maritime technology has been formally forbidden by Congress, grossly inefficient administration of the export control laws has allowed the Soviets to acquire a massive military transportation capacity.

Origins of the Soviet Merchant Marine

There are two extraordinary facts about the gigantic strategic Soviet merchant marine:

First: over two-thirds of its ship tonnage has been built outside the Soviet Union. The remaining one-third was built in Soviet yards and to a great extent with shipbuilding equipment from the West, particularly Finland and NATO allies, Great Britain and Germany.

Second: four-fifths of the main marine diesel engines used to propel the vessels of the Soviet merchant marine were actually built in the West. In other words, only one-fifth of the main diesel engines were built in the USSR. Moreover, even this startling statistic does not reflect the full nature of Soviet dependence on foreign marine diesel technology because **all of the main engines manufactured in the USSR are built to foreign designs.** The full scope of the dependence of Soviet marine-diesel technology on foreign assistance is shown in Table 8-1.

The manufacture of marine diesels in the Soviet Union has received considerable foreign technical assistance. Technical-assistance agreements were made with both M.A.N. and Sulzer in the 1920s, and the Soviet Union has continued since that time to receive M.A.N. (Maschinenfabrik Augsburg-Nurnberg) and Sulzer technology, in addition to assistance agreements with Burmeister & Wain of Denmark and Skoda of Czechoslovakia.

An important basic agreement was signed in early 1959 in Copenhagen by Niels Munck, managing director of Burmeister & Wain. The Danish company also has a licensing agreement with the Polish engine manufacturers Stocznia Gdanska, and most of that organization's annual production of B & W designs goes to the Soviet Union.

All modern large diesels of more than 11,000 horsepower used in the Soviet Union are built to a single foreign design — Burmeister & Wain of Copenhagen, Denmark. Denmark is a NATO ally of the United States. The export of this Danish technology could have been stopped by the State Department under the Battle Act and CoCom arrangement. All Burmeister & Wain diesels are designed with U.S. computers. Burmeister & Wain engines propelled the Soviet ships that were active in the Cuban Missile Crisis in 1962, the supply of North Vietnam and Central America today.

The State Department could also have intervened indirectly to restrict export of military technology by Eastern European Communist governments — for example, Skoda armaments — to the USSR and this was the very claim made by State to Congress in order to bring detente. In 1966 Dean Rusk submitted legislation to Congress for "most favored nation" treatment for East European Communist countries. This would, said Dean Rusk, "give the United States an important political tool in Eastern Europe." But East European Communist countries went right on providing technical assistance for the Soviets, and the State Department maintained a steadfast blind eye to the military end-uses of this technical assistance to the Soviet Union. Indeed, State even approved an important technical-assistance agreement made by Simmons, an American firm, with Skoda.

Table 8-1
Origin of Diesel Engines of Soviet Merchant Ships

Size of Merchant Ship (Gross Registered Tonnage)	Engines of Foreign Design and Construction (percent)	Engines Soviet-built under Foreign License (percent)
15,000 and over	100	0
10,000-14,999	87.9	12.1
5,000-9,999	56.9	43.1

*This includes diesel-electric units but not steam turbines. The chart is based on gross registered tonnage, not rated capacity of the engines, therefore it is an approximate measure only.

Illegal Actions by State Department

When we look closely at the transportation technology used to support the most dangerous international crises of the 60s, 70s and 80s, we find that the U.S. State Department not only had the knowledge and the capability to stop the technological transfers which generated the vehicles used, but was required by law to ensure that the technology was not passed to the Soviets. In other words, there would have been no Cuban Missile Crisis in 1962, no supply of the Vietnamese War and no wars of liberation in Africa and Central America if State Department had followed Congressional instructions and carried out the job it is paid to do.

The **Poltava** class of Soviet merchant vessels, equipped with special hatches, was used to carry missiles to Cuba in 1962, it was used to supply the Vietnamese War, it is used today to support "wars of liberation" in Africa and Central America. The main engines for the first two vessels in this class were manufactured by Burmeister & Wain in Copenhagen. Engines for the other ships in the class came from the Bryansk plant in the Soviet Union. The Danish and the Bryansk engines are built to the same specification: 740-millimeter cylinder diameter and 1,600-millimeter piston stroke. The Danish engines have six cylinders while the Soviet engines have seven cylinders; in all other respects they are identical Burmeister & Wain-design engines. In 1959 the Danish company made a technical-assistance agreement with the Soviets for manufacture of large marine diesels, **not manufactured in the USSR at that time,** and the U.S. State Department, through CoCom, approved the export of this technology as nonstrategic. Any member of CoCom has veto power. Objection by State Department representatives would have effectively blocked the agreement.

The **Poltava**-class ships were used to carry Soviet missiles to Cuba in 1962. The first **Poltava** engines were manufactured in Denmark in 1959 and the ships entered service in 1962, only a few months before they were used to transport missiles to Cuba. In other words, the first operational use of these diesel engines — approved by State as nonstrategic — was in a challenge to the United States which brought us to the brink of nuclear war.

The **Poltava**-class ships have extra long hatches: eight of 13.6 meters length and 6.2 meters width, ideal for loading medium-range missiles. After near nuclear conflict between the United States and the Soviet Union the Soviets removed their missiles — as deck cargo on other merchant ships. The **Labinsk** was one of the ships used. The **Labinsk** is a 9,820-ton freighter built in 1960 in Poland on Soviet account and has Italian engines, made by Fiat in Italy (8,000 bhp, eight cylinders, 750-mm. cylinder diameter, 1,3200-mm. piston stroke). This is the same Fiat Company that later in the 60s and 70s provided technical assistance for the largest automobile plant in the USSR.

In 1967, while the Johnson Administration was campaigning for yet more "peaceful trade" with the Soviet Union, Soviet ships previously supplied by our allies as "peaceful trade" were carrying weapons to Haiphong to kill Americans (see Table 8-2).

In addition to the ships listed in Table 8-2, the **Kuibyshev,** a 6,000-ton freighter built in the United States, the **Sovetsk,** built in Poland with Swiss engines, and the **Ustilug,** a 4,400-ton freighter with West German M.A.N. engines, have also been identified.

As Table 8-3 shows, if the State Department had done an effective job according to the laws passed by Congress, thirty-seven of the ninety-six ships would **not** have been in Soviet hands — and would not have been able to take weapons and supplies to Haiphong.

Table 8-2
Analysis of some Soviet Ships used on Haiphong Run

Soviet Registration No.	Year of Construction	Name and GRT of Ship	Place of Construction	
			Engines	Hill
M26121	1960	**Kura** (4,084 tons)	West Germany	West Germany
M25151	1962	**Simferopol** (9,344 tons)	Poland	Switzerland
M11647	1936	**Arlika** (2,900 tons)	United Kingdom	United Kingdom
M17082	1962	**Sinegorsk** (3,330 tons)	Finland	Sweden
M3017	1961	**Ingur** (4,084 tons)	West Germany	West Germany
M26893	1952	**Inman** (3,455 tons)	East Germany	West Germany

Specifically, the State Department could have stopped the export of marine-diesel technology to the Soviets under the Battle Act.

Could the Soviets have used other ships? Turn to the next page.

Over two-thirds of Soviet merchant ships and more than four-fifths of the marine diesels in Soviet merchant ships were **not** built in the USSR. The Soviets would certainly not have attempted foreign adventures with a merchant marine substantially smaller than the one they have now in operation. In other words, we always had the absolute means to stop the Soviet tide of aggression — if that was our objective.

The provision of fast, large ships for Soviet supply of the North Vietnamese indicates where export control failed.

Table 8-3
Engines of Soviet Ships on Haiphong Run and Ability of United States to Stop Export under Battle Act and CoCom

Origin of Diesel Engines	Number of Engines Manufactured In USSR	Number of Engines Manufactured Outside USSR	Could Export Have Been Stopped?
Manufactured in USSR to Soviet Design		—	—
Manufactured in USSR under license and to foreign design:			
Skoda (at Russky Diesel)	5		No
Burmeister & Wain (at Bryansk)		8	Yes
Manufactured outside USSR to foreign design:	8		
Skoda (Czechoslovakia)		5	No
M.A.N. (West Germany)		11	11 Yes
Fiat S.A. (Italy)		2	2 Yes
Burmeister & Wain (in Denmark and elsewhere under license)		8	8 Yes
Sulzer (Switzerland)		13	No
Lang (Hungry)		4	No
Gorlitz (East Germany)		10	No
United States (Lend-Lease)		7	7† Yes(?)
United States (not Lend-Lease)		1	No
Krupp (Germany)		1	1 Yes
Total: Diesel engines	13	62	37 Yes

†Lend-Lease — should be returned under the Master agreement.

Steam Turbines and Riciprocating Steam Engines	Number of Engines Manufactured		Could Export Have Been Stopped?
	In USSR	Outside USSR	
Manufactured in USSR to Soviet design	0		
Manufactured in USSR to foreign design	1 (possibly)		
Manufactured outside USSR:			
Canada		1	
United States		3	
United Kingdom		1	
Switzerland (Sulzer)		3	
Total: Steam turbines	1	8	

Grand Total:				
	Diesel engines	75	Not identified	12
	Steam turbines	9	Identified	84
		84		96

Segments of the Soviet merchant marine were examined to determine the relationship between Western origins and the maximum speed of Soviet ships. It was anticipated that because of NATO limitations on the speed of merchant ships supplied to the USSR (reflected in the export control laws) that the average speed of NATO-supplied ships would be considerably **less** than ships supplied either by Eastern European countries to the USSR or built within the USSR itself. The results of an analysis of forty-two Soviet ships on the Haiphong supply are as follows:

Merchant ships with engines manufactured in the Free World, average speed 14.62 knots.

Merchant ships with engines manufactured in Eastern Europe, average speed 13.25 knots.

Merchant ships with engines manufactured in Soviet Union, average speed 12.23 knots.

(All forty-two ships were built after 1951, the year the Battle Act was implemented.)

The most obvious point to be made is that the average speed of Western-supplied ships used by the Soviets on the Haiphong run was 2.4 knots (i.e., about 20 percent) above that of Soviet domestic-built ships on the run. This includes only those ships built after 1951 (i.e., after implementation of the Battle Act and its limitation of speed and tonnage on ships supplied to the USSR).

The illegal administration of the Battle Act by State Department also applies to weight limitations — the faster, larger Soviet ships are from the West and the slower, smaller ships are from domestic Soviet shipyards.

Under the CoCom machinery each nation participating in the embargo of strategic materials submits its own views on the shipment of specific items. There is also a unanimity rule. In other words, no item may be shipped to the USSR unless all participating nations agree. Objections by any nation halt shipment. Douglas Dillon, former Under Secretary of State, has pointed out, "I can recall no instance in which a country shipped a strategic item to the Soviet bloc against the disapproving vote of a participating member of CoCom."

It must therefore be presumed that the United States delegates approved the export of ships of high average speed, as well as marine diesel engines and the Burmeister & Wain technical-assistance agreement of 1959 for Soviet manufacturers of marine diesels, all of which were later used against the United States by the Soviets in supplying North Vietnam, and adventures in Africa and Central America.

It is clear from this evidence alone that successive administrations have been long on words but short on action to stop the Soviets from carrying out world ambitions. Further, successive administrations have committed American soldiers to foreign wars without the resolve to win and obviously in the knowledge that American technical assistance was being provided to both sides in these wars.

Under CoCom arrangements in the Battle Act, the State Department can enter an objection to CoCom concerning any technological exports to the Soviet Union. No CoCom member can make such exports over the objection of any other member of CoCom. In other words, if the State Department had **wanted** to implement the intent of Congress, it had the ability to stop the transfer of marine engines and marine engine technology to the USSR. It did not do so.

Even further, other Soviet ships have marine diesels originating in Czechoslovakia (Skoda), Hungary (Lang), and East Germany (Gorlitzer), countries for which State has demanded most favored nation status and trade as a "political weapon." If there is indeed a polycentralist trend, then why was the State Department unable to stop the flow of military technology to the Soviet Union? It had the political weapon (trade) it asked for to do the job.

In brief: State Department had the means to stop the transfer of marine-diesel technology. The department was required to do so under the law. It did not do so. The blame in this tragic case is squarely at the door of the State Department.

We can derive two conclusions:

1. The Soviet Union could not have supplied wars of liberation without assistance from the United States and its Western allies. This assistance takes the form of technology transferred through the vehicle of trade.

2. The State Department had the absolute means to stop this transfer through its veto power in CoCom. It did not do so.

The position is more serious than even these conclusions would suggest, because the State Department has excellent — and expensive — intelligence facilities. The department was therefore aware of item (1) above. It is also aware of its powers in CoCom. Yet a departmental spokesman went before Congress to make the following statement:

If there were no trade at all between West and East, the Soviet Union would still be perfectly competent to supply North Vietnam with its requirements, many times over. I think the proposition that our restrictions or any restrictions on trade with Eastern Europe can defer or affect in any significant way the ability to supply North Vietnam is simply wrong.

This bland assertion, without evidence, of course, was made to the Senate by Philip H. Trezise, former Assistant Secretary for Economic Affairs in the department. Trezise has been described by Senator Mondale as "one of the most remarkable men that this Committee could hear from." That is certainly an accurate statement. Unfortunately, Trezise has all the qualities of a deaf mute blindman.

The Deaf Mute Blindmen Forge Ahead

These facts and conclusions were evident by the early 1970s: the Soviets were using Western transportation technology against the West.

Yet in 1973 the State Department under President Nixon and National Security adviser Henry Kissinger negotiated a major agreement to transfer even **more** transportation technology to the Soviet Union. This agreement was signed by President Nixon on June 19, 1973 and went far beyond maritime technology. The maritime component was as follows:

Article 2

"Marine transport, including technology of maritime shipping and cargo handling in seaports."

This is, of course, the precise technology to supply for assisting Soviet wars of liberation.

We reproduce two pages from this extraordinary agreement (Russian and English versions) to demonstrate the all-encompassing nature of the Nixon Administration's aid to the Soviet Union — even when the results of earlier aid were known in Washington. When this "aid and comfort" is considered in the light of suppression of the facts, coupled with harassment of individuals attempting to bring these facts home to the American public, there is clearly a case for in-depth investigation of the motivation of top officials in the Nixon Administration, up to and including Mr. Nixon himself.

Submarine and Anti-Submarine Warfare

There is a long history of Western assistance to Soviet submarine construction and anti-submarine warfare efforts.

Extensive tsarist submarine work was adapted by the Soviets at the end of the 1920s and a few tsarist-model submarines were still operating in World War II.

Soviet domestic construction began in 1928 with the L and M classes. The L-class was based on the British **L-55**, which sank off Kronstadt and was raised by the Soviets; twenty-three of the L-class and one L-Special were built to this model by 1938. The M-class, a small 200-ton coastal submarine of limited performance, was made possible only by the introduction of electric welding under the terms of a General Electric technical-assistance contract.

All subsequent Soviet submarine development has been heavily influenced by German U-boat designs and more recently by U.S. designs. In 1926 a German naval mission under Admiral Spindler visited the USSR and provided the plans of the most successful German submarines, details of operational experience, and the services of German submarine experts. The Russians obtained sets of U-boat plans, the most important of which were those of the B-III type, one of the most successful designs for a conventional submarine ever produced. As the type-VII, the B-III was the backbone of the German U-boat fleet in World War II.

A variant of the design was built in Russia — first known as the N-class — nicknamed **Nemka** ("German girl") — and later as the S-class. The **Chuka**-class was based on German B-III plans; S-class (enlarged **Chuka**) is the German type VII U-boat.

Italian influence came in two submarine classes. Eight vessels in the **Garibaldi**-class were of Adriatico design and seventeen **Pravda**-class submarines were a development from the **Garibaldi**. Two submarines were bought from Vickers-Armstrong in the United Kingdom in 1936, and the Soviet V-class comprised Vickers-Armstrong submarines built in the United Kingdom in 1944 and transferred to the USSR under Lend-Lease.

The United States sold submarine equipment to the Soviet Union in the first five or six years of the 1930s. A proposal was received by the Electric Boat Company of Groton, Connecticut, in January 1930 for the construction of submarines and submarine ordnance equipment for shipment to the USSR. In a letter to the Secretary of State, Electric Boat

Sample Page From Transportation Agreement
Signed by President Nixon with the Soviet Union
June 19, 1973
[Russian Version]

Статья 2

Это сотрудничество будет направлено на изучение и решение конкретных проблем в области транспорта, представляющих взаимный интерес. На первом этапе сотрудничество будет осуществляться в следующих областях:

a) строительство мостов и тоннелей, включая проблемы контроля напряженного состояния и разрушения сооружений, специальные методы строительства в холодных климатических условиях;

b) железнодорожный транспорт, включая проблемы подвижного состава, пути, высокоскоростного движения, автоматизации и эксплуатации в условиях холодного климата;

c) гражданская авиация, включая проблемы повышения эффективности и безопасности;

d) морской транспорт, включая технологию морских перевозок и переработку грузов в портах;

e) автомобильный транспорт, включая проблемы безопасности движения.

Другие области сотрудничества могут быть добавлены по взаимному согласию.

Статья 3

Сотрудничество, предусмотренное в предыдущих статьях, может осуществляться в следующих формах:

a) обмен учеными и специалистами;

b) обмен научно-технической информацией и документацией;

c) проведение совместных конференций, совещаний и семинаров;

d) совместное планирование, разработка и осуществление научно-исследовательских программ и проектов.

Другие формы сотрудничества могут быть добавлены по взаимному согласию.

Sample Page From Transportation Agreement Signed by President Nixon with the Soviet Union June 19, 1973 [English Version]

ARTICLE 1

The Parties will develop and carry out cooperation in the field of transportation on the basis of mutual benefit, equality and reciprocity.

ARTICLE 2

This cooperation will be directed to the investigation and solution of specific problems of mutual interest in the field of transportation. Initially, cooperation will be implemented in the following areas:

a.   Construction of bridges and tunnels, including problems of control of structure stress and fracture, and special construction procedures under cold climatic conditions.

b.   Railway transport, including problems of rolling stock, track and roadbed, high speed traffic, automation, and cold weather operation.

c.   Civil aviation, including problems of increasing efficiency and safety.

d.   Marine transport, including technology of maritime shipping and cargo handling in seaports.

e.   Automobile transport, including problems of traffic safety.

Other areas of cooperation may be added by mutual agreement.

ARTICLE 3

Cooperation provided for in the preceding Articles may take the following forms:

argued that there was "no objection" to the construction of submarines for such "friendly foreign powers," and further said that this was in the interest of the Navy, as it kept domestic shipbuilders at work. The State Department, though admitting there was no legal restriction on shipments of munitions to the Soviet Union, said it viewed "with disfavor" the construction of periscopes, submarines, and ordnance equipment for shipment to the Russians.

There was also a flow of American technology under the Sperry Gyroscope technical-assistance contract for marine instruments, and many Soviet engineers were trained by the company in the United States, although attempts in 1937-38 to buy fire-control equipment were thwarted by Navy Department officers. By 1937 Electric Boat was negotiating with the Soviets for construction of submarines, this time with the blessing of the State Department.

The massive postwar expansion of the Soviet submarine fleet has depended upon the designs and technical and construction resources of Germany and the United States. After World War II the Soviets carefully studied German submarines and operational techniques. Using equipment and material received under Lend-Lease, and transferring complete shipyards and great quantities of submarine-building equipment from Germany, a large submarine construction program was undertaken — a program still in progress in 1986.

In 1972 the Soviet W-class attack submarine accounted for about half of the Soviet submarine fleet. The W-class is a direct copy of the successful German type-XXI U-boat. The vessel has a 1,621-ton displacement and is capable of traveling 11,000 miles without refueling. Although the Germans built 120 type-XXI boats by early 1945, few went to sea. Almost all of these completed submarines fell into Soviet hands. Thus a substantial portion, perhaps one-quarter, of the Soviet submarine force was built in Germany to German construction standards. A modification of type-XXI became the Soviet Z-class, slightly larger, with greater range. The most modern Soviet diesel-powered submarine, the F-class, was also developed from these advanced German designs.

Early Soviet nuclear-powered submarines are similar to the U.S.S. **Nautilus** in configuration. The Soviet Y ("Yankee")-class is copied from the U.S.S. **Polaris** ballistic missile submarine, with plans obtained through the massive Soviet espionage program in Great Britain.

Submersibles for deep-sea work have been purchased in the West, the most recent sale being the Hyco **Pisces-IV** sold in 1972. Missile-carrying submarines are fitted with GOLEM-class missiles. GOLEM I and GOLEM II are direct descendents of the German V-2, while GOLEM III is a two-stage solid-fuel equivalent of the Lockheed Polaris.

In anti-submarine warfare — an obvious priority for the Soviets — we find repeated efforts to obtain Western advanced anti-submarine technology, especially sensor technology, focusing on signal processing. This highlights the calculated planned nature of Soviet technology acquisition and the weak Western response.

Soviet military technology acquisition is focused upon 14 key technology clusters. These target groups are of fundamental importance to development of weapons systems. By concentrating on these fourteen groups, the Soviets try to acquire the broadest possible military advantages while simultaneously improving their own shortcomings.

One of these 14 groups includes sensor technology, comprising radar devices, array processors, infrared technology and signal processing. An example from one of these groups illustrates activity by U.S. firms which approaches treason.

Array processors assist a computer in processing and analyzing digital signals. This can be used to identify minute differences in sounds under the ocean, a means of locating enemy submarines. Obviously, if the Soviets have this technology, they can track U.S. Navy submarines.

In 1979 Geo Space Corporation of Houston, Texas sold 36 array processors to the Soviet Union. Soviet Navy personnel were **trained** in the Geo Space plant in Houston. Soviet personnel carried the Geo Space units aboard Soviet submarines and installed the units with the shipboard computers. This activity surely fits the definition of "aid and comfort" to enemies of the U.S. Yet Department of Commerce merely fined Geo Space $36,000 and suspended its export privileges.

When we compare the activity of Departments of State and Commerce in administering the Export Control and Battle Act laws we find extreme weakness, amounting in many cases to ignoring the will of Congress in favor of the Soviets.

Yet we also find another phenomenon, explored below.

The Soviet Union as a Source of Information

In practice the Soviet Union is a more prolific source of hard information on some technology transfers — information that can be blended with declassified U.S. files, and Congressional reports.

The availability of data on the origin of the main engines of Soviet ships used on the Haiphong supply run and in the Cuban Missile Crisis is a prime example of the Soviet Union publishing detailed information not available from U.S. government sources and directly conflicting with official statements made by U.S. government officials.

This case of the origin of Soviet vessels, for which ample and accurate hard data are available, is worth exploring. For most of the period since 1949, the Battle Act and the Export Control Act have **supposedly** prohibited the export of transportation technology for military purposes.

However, the specific case-by-case determinations made by State, Commerce, and CoCom within the framework of these laws are classified. It is not possible to obtain free access to the relevant decision papers to examine the manner in which the intent of Congress has been administered. We **do** know, however, that any member of the CoCom (Coordinating Committee, the operating arm of the Consultative Group established by NATO and Japan in 1950 to coordinate the export controls of the major industrial nations) group of nations has veto power and that no shipment has ever been made to the Soviet Union without the unanimous approval of all members. Thus, the transfer of Danish marine technology in 1959 had implicit or explicit U.S. State Department approval.

Some years ago research strongly suggested that the Soviets had no indigenous military transport technology: neither motor vehicles nor marine diesel engines. Yet about 80 percent of the weapons and supplies for the North Vietnamese were transported by some means from the Soviet Union. The greater part of these Soviet weapons went to Vietnam by Soviet freighter and then along the Ho Chi Minh trail on Soviet-built trucks.

By using data of Russian origin it is possible to make an accurate analysis of the origins of this equipment. It was found that all the main diesel and steam-turbine propulsion systems of the ninety-six Soviet ships on the Haiphong supply run that could be identified (i.e., eighty-four out of the ninety-six) originated in design or construction outside the USSR. We can conclude, therefore, that if the State and Commerce Departments, in the 1950s and 1960s, had consistently enforced the legislation passed by Congress in 1949, the Soviets would not have had the ability to supply the Vietnamese War — and 50,000 more Americans and countless Vietnamese would be alive today. The names of the ninety-six Soviet ships used on the Haiphong run were gleaned from *Morskoi Flot* and similar Russian maritime publications. The specifications of the main engines were obtained from *Registrovaya Kniga Morskikh Sudov Soyuza SSR* and other Russian sources. This hard information came from **censored** Soviet sources. The same information is only available in the West in classified government files; and it is therefore totally censored to the independent researcher and to Congress.

This is the paradox. The U.S. government is concealing, possibly unknowingly, actions which are aiding the Soviets, and which originate within U.S. government offices.

The Russia No. 6 Project

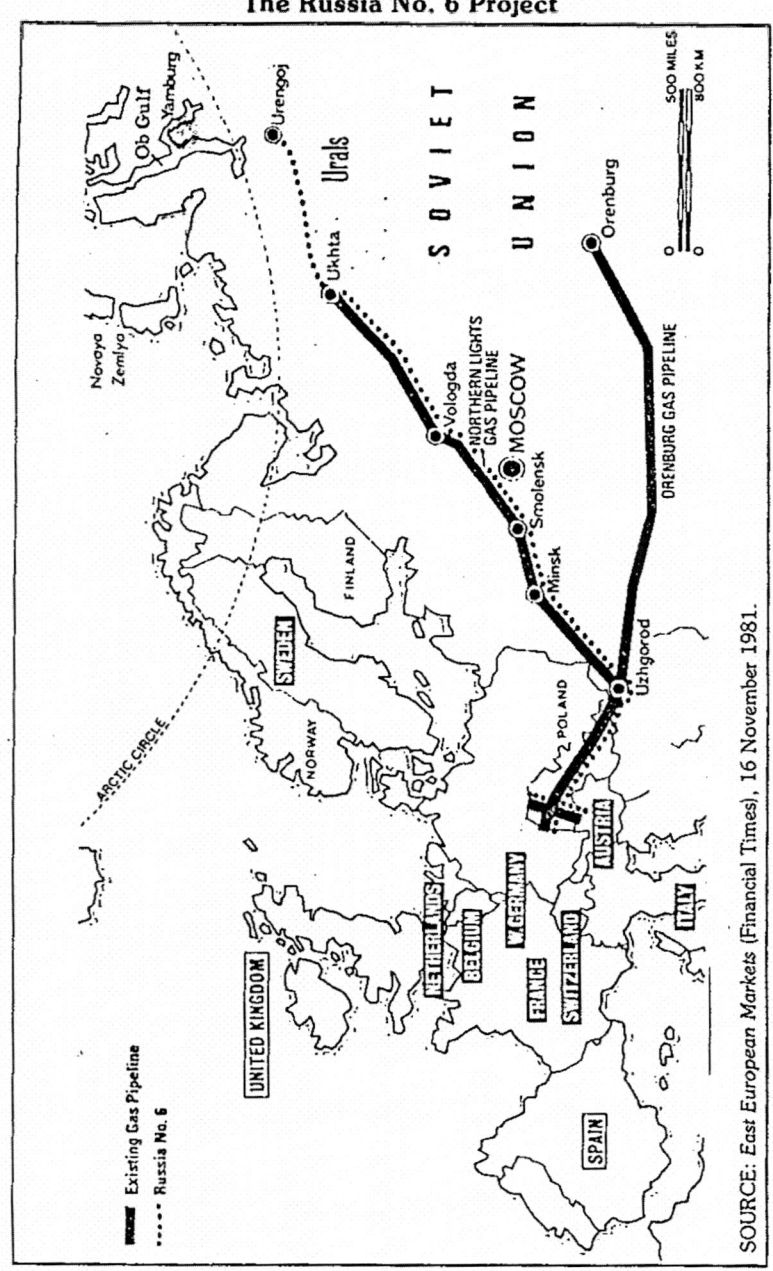

SOURCE: *East European Markets* (Financial Times), 16 November 1981.

SOURCE: Charles Levinson VODKA-COLA

The Vodka-Cola Web Surrounding
The Urengoi Pipeline

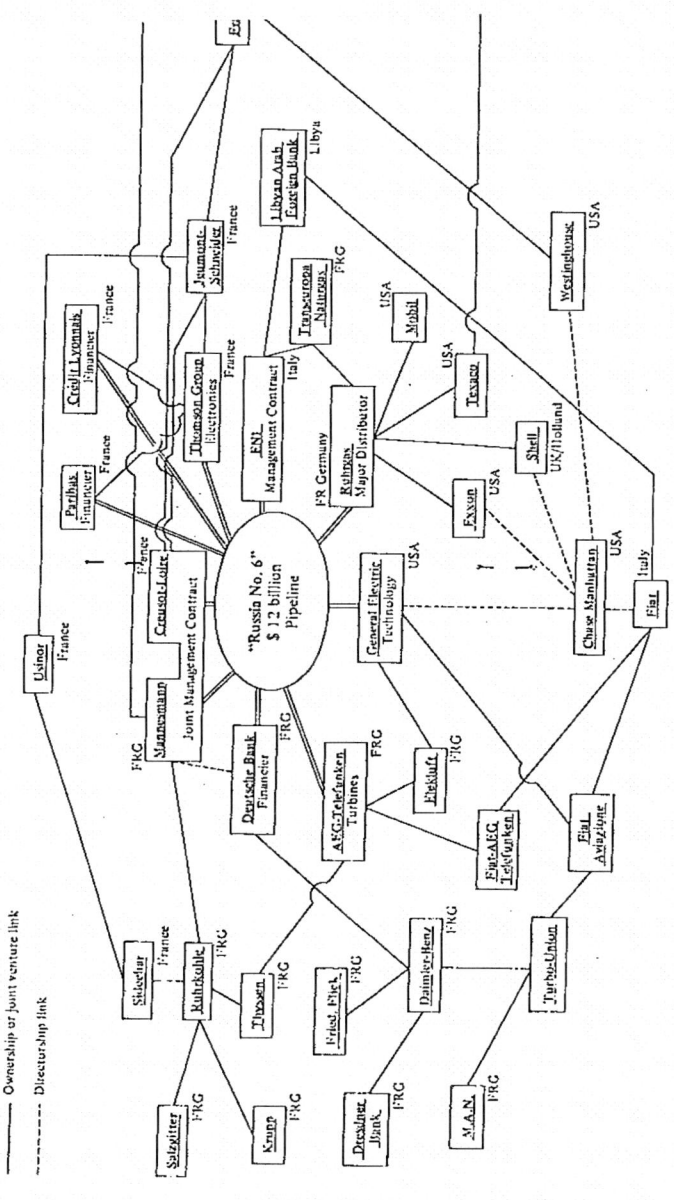

═══ Direct Pipeline Contract

─── Ownership or joint venture link

----- Directorship link

—131—

CHAPTER IX

The Leaky Pipeline Embargo[27]

"It seems beyond doubt now that the pattern of Soviet-Cuban-East German intervention constitutes a single apparatus with a serious purpose. It merits a serious American response."

Wall Street Journal, December 29, 1980

Siberia is a vast store house of oil and gas reserves. In the 1970s the Soviets brought off an almost unbelievable deal to develop these reserves.

A known Soviet objective in economic warfare is to make Western Europe dependent on the Soviet Union. Such dependence will severely reduce European options in case of war with the Soviet Union. The Soviets designed a massive pipeline project large enough to change the entire Siberian infrastructure to channel these Siberian natural gas reserves to gas deficient Europe, thus making Europe dependent on a vital energy resource. At the same time the Soviets convinced the Western deaf mute blindmen to finance this $22 billion deal and so finance their own destruction — just as Lenin predicted.

The Siberian gas deal, known as "Russia No. 6" to the financing bankers, is a 2,800 miles gas export pipeline from the Urengoy gas field in Siberia to Uzhgorod on the Czech-Soviet border, where it feeds into the West European gas pipeline network. Daily throughput is 2.8 billion cubic feet.

Initially, even the U.S. State Department (October 1982) objected to the deal on the following grounds:

• Russia No. 6 would make Europe 20%-30% dependent on Russian gas, thus crossing "the threshold of prudent dependence on the USSR,"

• the financing offered by Western bankers "amounts to a subsidy of Soviet economic development,"

• resulting hard currency earnings from sale of the gas will "have a strategic impact by allowing the USSR to continue to import Western goods and high technology equipment, alleviating serious domestic resource constraints."[28]

In the late summer of 1981, the Soviet Union contracted with U.S., French, West German, and Italian engineering firms for equipment for the Siberian pipeline. The Soviets required each European country to make available substantial export financing to be eligible for pipeline

[27]Much of this chapter is based on an excellent study, *The Vodka Cola Pipeline*, by Charles Levinson. Available from P.O. Box 277, 1211 Geneva 19, Switzerland. Send $6 per copy postpaid.

[28]U.S. Department of State, *Gist* "Siberian Gas Pipeline and U.S. Export Controls," October 1982.

contracts. Financing offers by the European export credit agencies were for subsidized interest rates, in most cases under 8%, at a time when interest rates ranged from 11% to more than 20%. Using credit, the Soviets purchased power turbines, gas compressors, and monitoring, firefighting, and control equipment. Large diameter pipe was purchased and financed as part of bulk orders of Western steel pipe regularly imported by the Soviets.

In late 1981 and early 1982, long-term gas supply contracts based on the new pipeline were signed between the Soviet Union and gas distribution companies in West Germany, France, Austria, Switzerland, and later Italy. The new contracts significantly increased West European dependence on the Soviet Union for natural gas, in some cases pushing it above 30% (see Table 9-1). Such a level of dependence on Siberian gas is unwise because:

- gas is a fuel particularly hard to replace at short notice because of the way it is transported and stored;

- gas is used in the politically sensitive residential and commercial sectors of Europe.

By the end of the present decade and after equipment credits are paid off, expanded gas exports will earn the USSR $8-$10 billion annually in hard currency. This will be a most important source of hard currency for the Soviets, offsetting an expected decline in crude oil export earnings.

Table 9-1
Gas Supply Sources for Major European Consumers

Federal Republic of Germany	1980	1990
Domestic	30.1%	22.3%
Netherlands	36.6	15.7
Soviet Union	17.5	25.0
Norway	15.7	22.3
Middle East	—	7.8
Africa	—	6.5

France	1980	1990
Domestic	27.5%	8.0%
Netherlands	37.6	8.0
Algeria	7.8	23.0
Soviet Union	13.3	32.0
Norway	9.1	12.0
W. Germany	4.0	—
W. Africa	—	16.0

Italy	1980	1990
Domestic	46.8%	17.0%
Netherlands	24.3	13.6
Algeria	—	27.2
Soviet Union	23.7	35.2
Libya	4.9	5.6

Source: *Financial Times* (London)

Working Both Sides of the Street

Financing of the Siberian gas pipeline is an excellent example of the two-faced nature of the deaf mute blindmen. In great part those who financed this vast expansion in Soviet ability to wage global war at Western taxpayers' expense are also prime military contractors for Western governments.

General Electric supplies guidance systems for Polaris and Poseidon missiles and jet engines for U.S. military aircraft, while at the same time supplied equipment for Soviet military end uses — on credit at preferential terms that could not be obtained by an individual U.S. taxpayer. In brief, because the U.S. government guarantees these Soviet orders, General Electric is in a position to have the U.S. taxpayer subsidize its contracts in the Soviet Union while the same taxpayer is shelling out for the U.S. Defense budget.

Here are the U.S. prime contractors on "Russia No. 6," with a brief notation of their U.S. government defense contracts:

Military Contracts with United States Government

General Electric Company
Total 1980 sales: $18,654 million; military sales: $2,202 million (No. 5 in USA)
• Produces jet engines for military aircraft including F-4 Phantom, F-5 Freedom fighter and F-18 Hornet.
• Guidance system for Polaris and Poseidon missiles.

General Electric Company
• Data processing systems.
• Components for nuclear weapons.

Exxon Corporation
Total 1980 sales: $63,896 million Military sales: $479 million (No. 29 in USA)

Royal Dutch/Shell Group
Total 1980 sale: $225,090 million
• Sales to U.S. DoD were $225 million in 1980, ranking it No. 52.

Contracts with Soviet Military End Use (In addition to "Russia No. 6")

General Electric Company
1981 — Romania ($142 mil.)
Steam turbine generating equipment for nuclear power station.
1980 — USSR ($40 mil.) Subcontract for computers and electrical equipment for electrical steel plant.

General Electric Company
1979 — Hungary
Licensed production of polypropylene film, also technology for condenser manufacture.
1979 — Yugoslavia
Know-how for manufacture of polyethylene cable.
1978 — Poland ($12 mil.)
Equipment for hot strip steel mill.
1976 — USSE ($90 million)
Hot gas rotating components.

Exxon Corporation
Participation, through holdings in Ruhrgas and Gasunie, in the distribution of Soviet gas in Europe.
1978 — Poland
Additives for and formulation of high quality lubricating oils.
1977 — USSR
Scientific and technical cooperation, exchange of petrochemicals, information and research cooperation in developing additives to lubricating oils, resins, solvents and semi-finished chemical products over 5 years.

Royal Dutch/Shell Group
1979 — Hungary
5-year cooperation in various field including marketing.

In 1978, Shell business with Hungary included $5 million in buy-back products.

1979 — Romania
Subsidiary General Atomic to supply Triga nuclear reactor for research center.

1978 — Bulgaria
Licensing of process for liquid plant protectant.

1978 — China
Licensing of process for $250 million methanol plant.

1976 — Poland
Licensing of process for ethylene oxide plant.

Ente Nazionale Idrocarburi (ENI) (U.S. Subsidiary is Agip)
Total 1980 sales: $27,186 million
• Subsidiary Agip had $223 million of business with U.S. DoD in 1989, ranking it 53rd among U.S. defense contractors.

Ente Nazionale Idrocarburi (ENI)
1981 — GDR
5-year cooperation agreement with Chemiaranlagen Export-Import.

1981 — China
Joint agreement for research and development of Chinese petrochemicals and synthesized polymers.

ENI (U.S. Subsidiary is Agip)

ENI (U.S. Subsidiary is Agip)
1981 — Romania
Framework agreement for complete production program of turbines, compressors, pumps, valves, etc.

1981 — Hungary ($3 million)
Supply and licensing of 30,000 tpa methyl terbutyl ether (MTBE) plant.

1980 — GDR ($90 million)
Snamprogetti: construction of highly advanced plant to recover lead from batteries.

1979 — China ($50 million)
Subsidiary Nouvo Pignone: joint production of centrifuge compressors.

1978 — Hungary ($80 mil.)
Supply of gas compressor station.
1976 — Poland
Subsidiary Haldor Topse: licensing of process for two ammonia plants.
1975 — USSE ($200 million)
Supply of three urea plants.

Foreign Companies with Significant Department of Defense Contracts

Rediffusion Ltd. (UK)

(Controlled by British Electric Traction Co. Ltd.)
• One of the top 10 suppliers of electronic equipment to UK military.
• R & D work for U.S. DoD (1977, $2.3 million); communications, data processing and flight simulation systems.

Rediffusion Ltd. (UK)

1981 — USSR
Supply and videotext systems and terminals for gas pipeline.
1979 — Czechoslovakia
($1.5 million) Supply of data transmission system and equipment.
1979 — Poland ($1 million)
Supply of data processing system and computers.
1978 — Czechoslovakia
($1.2 million) Supply of two computer systems.
1977 — USSR
Supply of computer system.
1977 — Czechoslovakia
($1 million) Supply of six data entry systems.
1977 — Poland ($1 million)
Supply of EDP computer equipment.

Thomson Group

Total 1981 sales: $8,656 million
• Variety of military electronics equipment including contracts on full range of Matra missiles such as Crotale, Martel, Otomot, and on MBB Kormoran antiship missiles.
• Surface radars, avionics, military data processing from Brandt Armaments Division.

Thomson Group

1979 — USSR ($100 million)
Supply of computerized telephone exchange system.
1979 — USSR
Construction of printed circuit plant at Minsk.
1979 — USSR
Data processing for nuclear plants.
1976 — Romania
Supply of air traffic control system.

Rolls-Royce
Total 1980 sales: $2,926 million; military sales: $250 million (No. 4 in UK)
• Propulsion systems for Bloodhound and Sea Dart missiles.
• Jet and turbine engines for military craft such as Harrier V/STOL fighter.
• With Turbomeca (France): Adour engines for Jaguar strike plane.
• With Detroit Diesel Allison: Spey engine for Corsair.
• With Fiat and MTU: RB199 engine for Panavia Tornado.

Fiat SpA
Total 1980 sales: $25,155 million
• Subsidiary Aeritalia (50%) builds military aircraft including F104 Starfighter. Partner in Panavia for production of Tornado strike plane.
• Subsidiary Sistemi Elettronica (SISTEL) builds Sea Killer, Indigo, Martel and Mariner missiles.
• Fiat Aviation Divsion builds turbine engines for military aircraft including General Electric licenses.
• In cooperation with MTU and Rolls-Royce: RB199 engine for Tornado.

1978 — Bulgaria
Agreement on electronics developments.

Rolls-Royce
1980 — USSR
Supply of "Avon" turbines for Surgut-Chelyabinsk pipeline.
1979 — Romania ($450 mil.)
Agreement for the supply of jet engines and eventual production.
1979 — China ($220 million)
License to produce supersonic Spey engines.
Licensed manufacture of aircraft engines.
1974 — Yugoslavia/Romania
RR Viper turbojets and afterburners for Yugoslav/Romanian Eagle jet fighter and Jastreb strike and tactical reconnaissance plane.
1975 — USSR
Scientific and technical cooperation in industrial motors.

Fiat SpA
1981 — USSR ($86 million)
Fiat-Allis to supply 300 63-ton heavy track loaders to USSR for use in large-scale civil engineering projects.
1981 — USSR ($90 million)
Earth moving machinery for strip mining in N. Siberia.
1980 — Hungary
Subsidiary Ercole Marelli: licensing of auto ignition tuner production.
1978 — USSR ($22 million)
Subsidiary Telettra to supply telecommunications network along Trans-Siberian railway.
1977 — Poland
Extension of licensing agreement for auto engines and automobiles. 260,000 vehicles produced under previous agreements.

1976 — USSR
Subsidiary Comau: subcontract for machine tools for nuclear energy components manufacturing plants (Comau sales to USSR for period 69-74 worth $100 million).

1977 — Hungary
5 year agreement for scientific and technical cooperation.

1976 — USSR
Agreement to expand auto input and for production of manufacturing, farm and building vehicles.

1976 — Bulgaria
General agreement on industrial and economic cooperation; joint R&D.

British Petroleum Co.
Total 1980 sales: $49,471 million
1977 sales to U.S. DoD worth $211 million.

British Petroleum Co.
1981 — China
Reconnaissance seismic survey in South Yellow Sea. First by any Western company.

1973 — USSR
Technology for 75,000 tpa chloroprene monomer plant in Yerevan.

1980 — USSR
Development of natural gas exploration and gas pipeline construction.

1977 — USSR
5-year exchange of technical information including refining lubricants, processes, and synthetic protein.

Foreign Companies with Siberian Pipeline Contracts and Significant European Defense Contracts

Friedrich Krupp GmbH
Total 1980 sales: $7,962 million
• Subsidiary of AG Weser: military ships for West German Navy.
• Subsidiary Mak Maschineenbau GmbH: tanks for West German Army; also development of a remote controlled mine-sweeping system.
• Subsidiary Krupp Atlas Electronik: electronic equipment including simulators and sonar systems.

Friedrich Krupp GmbH
1981-82 — USSR
Supply of steel pipe.
1980 — GDR ($875 million)
Construction of steel mill.
1980 — Poland ($137 mil.)
Construction of coal gasification plant.
1979 — USSR ($192 million)
Construction of electric steel plant.
1977 — USSE ($82 million)
Construction of DMT plant.
1977 — Hungary ($7 mil.)
Vegetable oil processing plant.
1977 — Czechoslovakia
Supply of effluent treatment plant; tire-making machine.
1979 — Poland
Cooperating in engineering and plant construction, raw materials, construction materials and foods.
1975 — GDR
Long-term economic and technical cooperation deal including 3rd market sales.

Creusot-Loire
Total 1981 sales: $3,805 million
• General steel products (gun turret mountings, etc.), engines, military vehicles.

Creusot-Loire
1980 — USSR ($300 million)
Prime contractor for steel complex at Nololipetsk.
1979 — GDR ($350 million)
Construction of nitrogen fertilizer plant, cooperation in sales.
1978 — USSR
Supply of turnkey petroleum coke calcination unit at Krasnorodsk.
1975 — USSR ($500 million)
Supply of chemical and ammonia plants, all paid for on buy-back basis.
1975 — USSR ($37 million)
Supply of equipment for Orenburg natural gas complex.

1979 — Bulgaria
Cooperation in mechanical engineering, chemicals, power engineering.

Aeg-Telefunken
Total 1980 sales: $6,750 million; military sales: $755 million (No. 2 in Germany)
• Guidance system on Euromissiles Roland, Hot and Milan.
• Subcontracts for avionics and electrical equipment on Tornado fighter plane.
• Electronic defense equipment including radar and navigation systems.
• TR86 computer for DISTEL command and control system used by NATO and USAF.

Aeg-Telefunken
1981 — GDR ($40 million)
Subcontract to supply heavy electrical equipment for plate rolling mill at Eisenhuttenstadt.
1979 — USSR ($22 million)
Telecommunications network.
1979 — China ($22 million)
Electrical equipment for three chemical plants.
1979 — Romania ($1.4 mil.)
Sale of Thyristor rectifiers for acetylene extraction.
1979 — Poland
Motors for power trucks and utility vehicles.
1977 — USSR
In cooperation with Mannesmann AG (FRG) built a section of Orenburg natural gas pipeline ($875 million in total).
1977 — Bulgaria
Contract to plan equipment and supervise expansion of electrical engineering works.
1978 — Yugoslavia
Joint production of copying machines and spares.

There is a simple, direct message in the above listings. Some of the most famous multinational corporations are sufficiently amoral to accept military contracts from both sides. This is not a new story. It occurs time and again in the history of the past century, but rarely has it been possible to identify the double-dealing while in progress.

While financing highly strategic projects for the Soviets, these multinationals are selling weapons and supplies to Western governments. An obvious deduction is that these corporations have little incentive to reduce world tension. **They maximize profit by keeping both sides in a state of near conflict.**

As Charles Levinson, author of *The Vodka Cola Pipeline*, phrases it:[29]

> Western workers must now be asking themselves [some] questions: why should they, through their taxes, bail out the banks who lent money to the enemy? Who gave Deutsche Bank permission to lend cheap money to Leonid Brezhnev, freeing him to divert domestic resources to the huge standing Soviet Army, Eastern Europe's "enforcer." Under what authority do Exxon and Shell, working through layers of overseas subsidiaries and holding companies, push to increase Western European dependence on Soviet energy sources?
>
> Despite the power of the interests at work, the system is not yet concretised and may yet be reversible. But far more public examination and debate is needed. And as time goes on, the ability of the power elites to keep down protest on both sides grows greater. Recent experience cannot be allowed to become a precedent: we cannot accept that human rights are worth less than a pipeline.

The Reagan Administration Marshmallow Approach

The initial public reaction of the Reagan Administration to Russia No. 6 was realistic enough. State Department issued a two page summary listing the dangers to world stability and Western freedom. Part of this report is reproduced on the first page of this chapter.

Then came months of heated debate within the Administration and heavy outside pressure from the deaf mute blindmen on the Reagan Administration and Congress. In general, conservative members of the Administration and Department of Defense made strong arguments against U.S. technology or exports for the Russia No. 6 project. In particular, Secretary of Defense Casper Weinberger and DoD Assistant Secretary of International Security Richard Perle made strong public arguments urging a widespread clampdown on U.S. subsidy of Soviet economic development. They argued correctly that we compounded our own defense problems with these projects.

On the other hand, leftovers from the Henry Kissinger-Richard Nixon detente school won the day. Led by then Secretary of State Alexander Haig (a protege of Henry Kissinger) and backed by Congressmen under pressure from the deaf mute blindmen pushed the Reagan Administration to collapse. In meeting with European allies Haig was accused of not pushing the Administration position and allowing the Europeans to continue with the deal by default. Without question the U.S. could have offered the Europeans an energy package of Alaskan oil, Canadian liquified gas and U.S. coal to replace Russia No. 6.

[29]*Op cit.*, p. 40.

In brief, Russia No. 6 demonstrates the ongoing political power of the deaf mute blindmen, in this case that group centered on David Rockefeller of Chase Manhattan. The U.S., even after six decades of subsidy of the Soviet economy and with a defense budget approaching $300 billion annually, was pressured into a project that gave the Soviets hard currency earnings, subsidised credit, and our finest technology — at U.S. taxpayers' expense.

CHAPTER X

DMBs Supply Nerve Gas Plants

"The USSR is highly dependent on Western chemical technology.
. . ."

Central Intelligence Agency

Chemical technology is an all-important prerequisite for modern warfare. Explosives require chemical technology and, for example, under wartime circumstances fertilizer plants can be quickly converted to manufacture of explosives. Many nerve gases require chemical technologies similar to those used for production of agricultural insecticides. This interrelationship between chemical technology and warfare is well known in Washington, yet the Soviets have traditionally been allowed access to the latest of Western chemical technology under so-called cooperation agreements, through "turn-key" plants which have been used for military end uses.

A Central Intelligence Agency assessment report[30] made in the late 1970s concluded that the "USSR is highly dependent on Western chemical technology." At that time the CIA estimated that Western equipped plants accounted for the following proportions of Soviet chemical production:

40 percent of complex fertilizers
60 percent of polyethylene
75-80 percent of polyester fiber
85 percent of ammonia production

The CIA report did not, however, report on another critical fact: that Soviet plants producing these and other chemicals use almost entirely technology copied or reverse engineered from Western equipment. There is no indigenous Soviet chemical technology.

State Department Concurs in Explosives Manufacture

Import of U.S. chemical technology for military purposes goes back to the 1920s and has always received State Department support.

A 1929 agreement stipulated that the Soviets could use Du Pont processes for the oxidation of ammonia to manufacture 50-65 percent nitric acid. Du Pont agreed "to place at the disposal of Chemstroi sufficient data, information and facts with respect to the design, construction and operation of such plants as will enable Chemstroi to design, construct and operate ammonia oxidation plants."

Later in 1932, negotiations were concluded between Du Pont and the Soviets for construction of a gigantic nitric acid plant with a capacity

---

[30]Central Intelligence Agency, National Foreign Assessment Center, *Soviet Chemical Equipment Purchases from the West: Impact on Production and Foreign Trade*, October 1978.

of 1,000 tons per day. This approximates 350,000 tons annually. Twenty-five years later, in 1957, the largest Du Pont-process nitric acid plant in the United States, at Hopewell, had an annual capacity of 425,000 tons. Under the 1929 earlier contract, Du Pont also supplied technical assistance to the USSR for a period of five years. The firm inquired of the State Department whether this plant of "excessively large capacity" would meet with objections from the U.S. government: "While we have no knowledge of the purpose of the proposed plant, yet the excessively large capacity contemplated leads us to believe that the purpose may be a military one."[31]

The State Department position is summarized in a memorandum of April 6, 1932, which reviewed export of military materials to the Soviet Union and concluded that the department would have no objection to construction of such a large nitric acid plant.[32]

The Soviets bought from Du Pont its ammonia oxidation and nitric acid technologies. Du Pont had expended over $27 million developing these processes. In requesting advice from the State Department, Du Pont argued that the process was neither secret nor covered by patents, that the end-use of nitric acid is the manufacture of fertilizer, that if Du Pont did not supply the process, it could be bought elsewhere, and that several plants had already been erected in the USSR by Casale and Nitrogen Engineering of New York.

The letter from Du Pont to Henry L. Stimson of the State Department with reference to the proposed contract (dated April 20, 1929) states in part, "It is true of course that nitric acid is used in the manufacture of munitions." Du Pont then claimed, "It is impossible to distinguish between chemicals used for strictly commercial purposes and chemicals used for strictly munitions purposes." And as justification for its proposal, the firm said, "We submit that the contemplated contract will in no way give assistance for the manufacture of munitions which cannot easily be acquired elsewhere by the Soviets."

Further, the company argued, there was nothing exclusive about the Du Pont process. The copy of the agreement in the State Department files indicates that the Soviet union

> [wishes] to use in Russia the Du Pont process for the oxidation of ammonia and [Du Pont] to place at its disposal sufficient data with respect to the design, construction and general information as to permit the satisfactory operation of such plants . . . the Company shall serve the Russian Corporation in an advisory capacity and furnish upon request services of engineers and chemists so as to accomplish the purpose of the contract.

[31]U.S. State Dept. Decimal File, 861.659 Du Pont de Nemours & Co/5, Du Pont to Secretary of State Stimson, Feb. 19, 1932.

[32]U.S. State Dept. Decimal File, 861.659, Du Pont de Nemours & Co/1-11.

By the late 1930s a Nitrogen Engineering-designed complex at Berezniki employed 25,000 workers and manufactured thermite, powder, and nitroglycerin.

In the late 1950s and 1960s, the Soviets lagged in all areas of chemical production outside the basic chemical technology absorbed in the 1930s and 1940s. This lag had major military implications and since 1958 has inspired a massive purchasing campaign in the West. In the three years 1959-61 alone, the Soviet Union purchased at least fifty complete chemical plants or equipment for such plants from non-Soviet sources. The American trade journal *Chemical Week* commented, with perhaps more accuracy than it realized, that the Soviet Union "behaves as if it had no chemical industry at all." Not only was Soviet industry producing little beyond basic heavy chemicals, but of greater consequence, it did not have the technical means of achieving substantial technical modernization and expansion in a product range essential for a modern military state.

Western firms supplied designs and specifications, process technology, engineering capability, equipment, and startup and training programs. These contracts were package deals that provided more than the typical Western "turn-key" contract. Such contracts are unusual in the West (except perhaps in underdeveloped areas lacking elementary skills and facilities) but were very attractive and highly profitable to Western firms.

Many of the chemical plants built in the 1960 and 1970s programs had direct military applications. In 1964 a British company — Power Gas Corporation, Ltd. — built a $14 million plant for the manufacture of acetic acid in the USSR. In 1973 Lummus Engineering built another vast acetic acid plant with Monsanto technology and U.S. Export-Import bank loans.

Hygrotherm Engineering, Ltd. of London contracted to supply an automatic heating and cooling plant (with heat generators, circulating pumps, and control equipment) and other equipment for use in the manufacture of synthetic resisn. A plant was supplied for the production of synthetic glycerin, which is used in explosives manufacture. Other plants were for the production of ethyl urea, synthetic fatty acids, sodium tripolyphosphate, carbon black, and germanium. All these products have military end-uses.

Sulfuric acid, the most important of inorganic acids and industrial chemicals, is required in large quantities for explosives manufacture. Production of sulfuric acid in Russia increased from 121,000 tons in 1913 to just under 3,000,000 tons in 1953, 4,804,000 tons in 1958, and 8,518,000 tons in 1965.

The Soviets have always utilized basic Western processes for the manufacture of their supply of sulfuric acid and have reverse engineered the equipment in their own machine-building plants. A recent Russian paper on sulfuric-acid manufacture indicates that in the mid-1960s, 63 percent of sulfuric-acid production was carried out according to a standardized version of one Western process. The remainder was produced by a "Soviet process" (utilizing fluidized bed roaster, electric precipator, towers, and contact apparatus) similar to the contact processes in use in the West. In 1965 Nordac, Ltd. of the United Kingdom sold a sulfuric-acid concentration plant with a capacity of 24 tons per day of 78 percent sulfuric acid to update Soviet sulfuric acid technology.

Up to 1960 Russian output of fertilizers was mostly in the form of low-quality straight fertilizers. There was no production of the concentrated and mixed fertilizers that are used in the West. Fertilizer plants are easily converted to explosives plants. Part of the fertilizer expansion program of the 1960s was the purchase from the Joy Manufacturing Company of Pittsburgh of $10 million worth of equipment for potash mining. Congressman Lipscomb protested the issue of a license for this sale (*Congressional Record*, Aug. 28, 1963). While Lipscomb pointed out that potash can be used for manufacture of explosives, Forrest D. Hockersmith, of the Office of Export Control in the Department of Commerce, replied, "Our decision to license was heavily weighed by the fact that potassium fertilizer can best be characterized as 'peaceful goods' " (Aug. 21, 1963). Hockersmith did not, of course, deny that potash had an explosives end-use.

A cluster of ten gigantic fertilizer plants for the Soviets was arranged by the Occidental Petroleum Corporation (Armand Hammer's company) and built by Woodall-Duckham Construction Company, Ltd., and Newton Chambers & Company, Ltd., of the United Kingdom. Other fertilizer plants were built by Mitsui of Japan and Montecatini of Italy. Ammonium nitrate, an ingredient in fertilizer manufacture, also has an alternate use in explosives manufacture. It is used, for example, in 60/40 Amatol in the explosive warheads of the T-7A rockets.

Armand Hammer of Occidental Petroleum is, of course, Moscow's favored deaf mute capitalist, possibly vying with David Rockefeller for the honor. However, Armand has a personal relationship with the Soviets that could never be achieved by anyone with David's Ivy League background. One fact never reported in U.S. newspaper biographies of Armand Hammer is that his father, Julius Hammer, was founder and early financier of the Communist Party USA in 1919. Elsewhere this author has reprinted documents backing this statement, and translations of letters from Lenin to Armand Hammer with the salutation "Dear Comrade."

That Armand Hammer and Occidental Petroleum would supply the Soviets with massive plants that can quickly be converted to explosives manufacture is no surprise. What **is** a surprise is that Armand Hammer has had free access to every President from Franklin D. Roosevelt to Ronald Reagan — and equal access to the leaders in the Kremlin.[33]

The DMB and Nerve Gas Technology

Chemical weapons are poisonous gases used to kill or incapacitate with toxic fumes either through the respiratory system or the skin.

These gases are deadly in extremely small amounts, only one milligram of Sarin can be lethal. Their action is to inhibit the enzyme action and inhibit normal body functions. They are extremely effective, causing death within minutes or more painfully over several hours.

The most potent — supertoxic lethal — chemicals are nerve gases first developed just before World War II, never used in warfare and now stockpiled by the United States, the Soviet Union and France. These chemical agents were derived from research in insecticides.

The major nerve gases are the G agents, Sarin (GB), Soman, Tabun, and the V agents such as VX. The original agent, Tabun, was discovered in Germany in 1936 in the process of work on organophosphorus insecticides. Next came Sarin and Soman and finally the most toxic, VS, a product of commercial insecticide laboratories.

The Soviet Union produced chemical weapons during the 1920s by cooperating with Germany. During World War II the Soviet Union produced considerable amounts of chemical weapons. Soviet stocks were not destroyed after the war and Soviet production continues. According to the U.S. representative at the United Nations in 1981, the Soviet Union maintained and operated at least 14 chemical weapons production facilities. Soviet chemical weapons production takes place in sealed-off sections of "civilian" chemical complexes, most built by Western companies.

Soviet stocks consist of the nerve gas Soman, a mustard gas, hydrogen cyanide, and other chemical agents from World War II, e.g., phosgene, adamsite and tabun. Estimates of the total stockpile vary from about the same amount as U.S. chemical agents to three times greater.

There is no distinction therefore between commercial chemicals such as fertilizers and chemicals used for war. In fact, a fertilizer industry can, with little trouble, be converted into an explosive industry and insecticides are the basis for chemical weapons including nerve gases.

There are other direct links between the nerve gases and commercial production. The very close chemical and toxicological similarity of organophosphorus pesticides to the nerve agents has made commercial

[33]The Armand Hammer story for the recent decades is covered in Red Carpet, cited on page one.

pesticides itself a major field of investigation regarding chemical weapons.

The best known example has been Agent Orange, a mixture of two herbicides, 2,4,5-T and 2,4-D, used in Vietnam. In the process of manufacturing trichlorophenol, an intermediate of 2,4,5-T, the highly toxic dioxin (TCDD) is produced in a side reaction. Dioxin is the same deadly chemical responsible for the disastrous accident at Seveso, Italy, where trichlorophenol was also manufactured as a feedstock.

While the nerve gases are the most lethal and dangerous chemical warfare agents, other lethal chemicals also exist, including those which caused the 1.3 million casualties of gas warfare during World War I. The major gases used were chlorine, phosgene and sulphur mustard. Other chemicals such as hydrogen cyanide and cyanogen chloride, which also have high toxicity levels, proved to be too uncontrollable for efficient use during the first war. These chemicals attack the lungs or the blood cells. They can cause permanent health damage, but to induce death must be deployed in much larger doses than the nerve agents. With the exception of mustard gas these are deceptive "dual purpose agents," widely used in industry for peaceful purposes, but capable of becoming weapons. Widespread use in the civilian chemical industry means that they are perhaps the most readily available chemical warfare agents for a country seeking a quick chemical arms stockpile.

Phosgene (Carbonyl Chloride, $COCL_2$) is the most obvious example. 50,000 tons of this highly toxic chemical were used for weapons during World War I. Today millions of tons are produced for dyestuffs and polyurethane building materials. **Hydrogen cyanide** (Hydrocyanic acid, HCN) is used for chemical fibers and produced in large amounts.

Unbelievably, Western companies have supplied not only "turn-key" plants, but the latest in information under long-term agreements. Here's a list of projects for the period 1975 to 1985.

Western Technical Assistance for Soviet Plants with Chemical Warfare Capability — 1975 to 1985

1975 to 1985	Schering AG, West Germany: license agreement License for production of the weed-killer betanal, up to 2,000 tons per year.
1976 to 1986	Rhone-Poulenc S.A., France: major cooperation agreement
	— agreement (1) sale to USSR of technology and equipment to produce pesticides and fertilizers to FF 2.5 billion; (2) repayment by chemicals from Soviet Union, 1980-1990, (3) sale to USSR of Rhone-Poulenc products to FF 1 billion.

— Financing equipment through FF 12 billion credit line. Buyback includes 30,000 tons per year of methanol and 10,000 tons per year of orthoxylene.

— Supply of 2,000 tons per year plant at Sumgait to produce Rhone-Poulenc process **lindane** insecticide. Value FF 176 million, with construction and know-how from Krebs-France.

— Supply of 10,000 tpa plant at Navoi to produce Rhone-Poulenc **organophosphorous insecticide "Phosalone."** Value $100 million, all equipment from France, with construction and know-how by the French engineering firm Speichem.

1976 to 1985	Monsanto, U.S.A.: technology Monsanto to provide process to make raw material **thonitrophenal** for the Navoi plant.
1978 to date	Sandoz, Switzerland: license Agreement on cooperation in agricultural chemicals, Soviets have obtained six licenses for **Sandoz pesticides,** and more to be tested.
1980	Iskra Industries, Japan: cooperation agreement Joint R&D to develop plant protection agents, **pesticides** and seed treatment agents.
1980 to 1990	Rhone-Poulenc, France: long term cooperation agreement 10-year FF 30-40 billion agreement for trade in products and technology. Rhone-Poulenc to supply turnkey agrochemical plants, agrochemical products, chemicals, fertilizers and animal feed in exchange for USSR naptha, ammonia, methanol and crude oil.
1981 to 1985	ICI, United Kingdom: trade agreement Agreement to trade L 63 million annually by 1985, including dyes, pigments, **pesticides, insecticides,** plastics, silicon compounds and surface active agents. USSR to supply bulk chemicals to ICI including acids, glycols, ammonia and methanol.
1981 to 1991	Rohm and Haas, USA; cooperation agreement Renewed for 10 years, an agreement for cooperation with USSR in areas of petrochemicals, plastics, agrochemicals, ion-exchange resons. Exchange of information, seminars and symposia, with joint testing of products.

1981 to	Lurgi-Gruppe, West Germany
1986	DM 370 million contract to construct plant in Volgograd by 1986 for a wide variety of **insecticides**. Lurgi has built a similar plant for the Soviets in Tashkent.
1981	Montedison (through Technimont), Italy Montedison negotiated sale of L 100 billion pesticide plant in exchange for raw materials.
1981	Sogo, France Soyuzchimexport order for cotton plant protection agents.
1981	Stauffer, USA Stauffer to deliver 11,000 tons of **herbicides** from Belgium to the Soviet Union.

What we are faced with is obviously an unholy cooperation between capitalists and communists to develop and produce dangerous weapons of unusual barbarity. The Soviets have used chemical weapons in Afghanistan and through client states in the Far East. One would expect a storm of protest from Western environmental groups. So far, the silence is deafening. Only the International Federation of Chemical, Energy and General Workers Unions has raised its voice in protest (See Bibliography). The national and locals in the U.S. are slow to follow the lead of the international — and this is the **only** voice of protest.

Two Russian Airmen Captured in Angola

SOURCE: NOW January 16, 1981

CHAPTER XI

Chevron-Gulf Keeps Marxist Angola Afloat

"Now, in spite of the increasingly advantageous position of UNITA, and the imminent collapse of the illegitimate, pro-Soviet government, elements within the State Department are doing their best to salvage the (communist) MPLA, and to prevent the forces for democratic government from winning in Angola."

Senator Steve Symms, May 14, 1985

In distant, obscure Angola, on the southwest coast of Africa, the alliance between capitalists and communists has matured into its most open blatant form. Possibly the reported export of ballbearing machines to Russia to produce bearings for Soviet missiles may leave some readers with an uneasy feeling, but also with a vague need to come to grips more concretely with the alliance. In Angola the most skeptical reader can see the alliance working on a daily basis.

Angola, a one time Portuguese colony, was "liberated" ten years ago by the MPLA (Popular Movement for the Liberation of Angola), a Marxist organization in alliance with Angolian nationalist groups. The MPLA was not elected and has never held elections. MPLA seized power and has been kept in power by 36,000 Cubans and about 1,200 Soviet military personnel. The Cubans and the Soviets are in Angola for the same reason Angolan Marxists will not allow free elections: because Marxism does not represent the people of Angola.

The United States has been inhibited from encouraging democratic, freely-elected forces by groups within the U.S. State Department and the so-called Clark Amendment (repealed in 1985), sponsored by former Senator Clark, which forbade the U.S. from assisting any group that might challenge the Angolan Marxists.

Yet the Soviets in Angola are challenged as they have never been challenged before; for the first time in 60 years a Marxist regime is in danger of overthrow by internal democratic forces. Angolan guerrillas, known as UNITA (National Union for the Total Liberation of Angola) have taken over one-third of Angola, about 250,000 square miles, and control most of the countryside, particularly in the south.

UNITA is an unusual organization. It is not American-backed. In fact, it is American corporations and the U.S. State Department that have stopped a UNITA victory. UNITA is unusual also in that it believes in free enterprise, free and secret elections, private property, and decentralization of political power. UNITA is led by Jonas Savimbi, aged 51, a ferocious-looking gentleman reminiscent of television's "Mr. T." Savimbi is a European educated black intellectual who believes in individual freedom.

Against Savimbi and UNITA we find the Soviet Union, Cuban forces, the U.S. State Department, American multinationals, and until recently, the U.S. Congress. Some years ago the Senate passed the Clark Amendment, sponsored by Senator Clark, which in effect prevented U.S. aid to this torch of freedom in southern Africa.

The muddled, confused thinking of the United States is well illustrated by a statement made by former U.S. Ambassador to United Nations Donald McHenry, to the effect that the U.S. should not be surprised that the Soviets are aiding Angolan Marxists: "That the Soviets are present to assist the Angolans and to assist the Namibians . . . is no different from the presence of the United States in El Salvador and U.S. assistance to El Salvador."

The point, of course, that McHenry avoids is that the Soviet objective, more than clearly demonstrated in the past 60 years, is a totalitarian controlled society without individual freedom.

The Soviets have indeed assisted Marxist Angola. As far back as 1981 Soviet military officers up to the rank of Colonel were killed and captured in Angola. Soviet Air Force personnel have been captured in Angola (see *Now*, January 16, 1981.)

The real oddity in Angola is that the single most important factor preventing a free open society is an American multinational corporation. As succinctly stated by Congressman William L. Dickinson (July 1985), "These Cuban troops are protecting American oil interests and they are preventing UNITA from overrunning the MPLA."

In northeast Angola is the Cabinda oil complex owned by Gulf Oil Corporation (since March 1984, part of Chevron Oil of California). CABINDA PROVIDES AT LEAST 80 PERCENT OF MARXIST ANGOLA'S FOREIGN EXCHANGE. The balance comes from diamond concessions operated by Anglo-American Corporation. Soviet and Cuban assistance is paid for from these foreign exchange earnings.

When we look closely at Chevron Gulf, we find that no less than a former U.S. Secretary of Defense, David Packard, has been in a position to thwart Gulf backing for Soviet Angola — yet did nothing.

Gulf Oil Corporation owns Cabinda, and Gulf itself was taken over by Chevron in March 1984. Thus, we have two sets of directors to look at, the original Gulf Oil directors who for a decade allowed the Gulf Cabinda operation to finance Marxist Angola, and the Chevron directors who had the opportunity to change corporate policy towards subsidy of Marxist warfare.

The directors of the former Gulf Oil Corporation were:

Jerry McAfee   Robert Dickey, III
H.H. Hammer   Julian Goodman
R.H. Dean   Sister Jane Scully
J.H. Higgins   Edwin Singer
J.P. Gordon   E.B. Walker, III
J.E. Lee   J.M. Walton
        E.I. Colodny

Of these the most vocal in support of Marxism was James E. Lee, former Chairman and Chief Executive officer of Gulf and now a director of Chevron. Lee was strong in support of Marxist Angola, even claiming to the *Wall Street Journal* that the Neto regime was "stable" and "easy to work with." (see cartoon opposite title page.)

In March 1984 Gulf was taken over by Chevron in the largest corporate merger in U.S. history. A few Gulf directors joined the Chevron board and Chevron-Gulf continued to operate Gulf Cabinda, protected by Cuban and Soviet troops, continued to provide most of Angola's foreign exchange and with the Angolan government, planned new joint ventures to expand corporate usefulness to the unelected Marxist government. That Chevron-Gulf should be protected by Cuban troops with Soviet air cover and a Soviet air defense network doesn't seem to embarrass these Chevron directors at all, **even though some are directors of major U.S. defense contractors:**

Samuel H. Armacost (45), Pres., Dir. & Chief Exec. Off. of Bank of America NT&SA.

Donald L. Bower (61), Vice-Chmn. of Bd. of Co.; Dir., Crocker National Corp., Crocker National Bank.

R. Hal Dean (68), Dir., Ralston Purina Co., Gulf Corp., Mercantile Trust Co., Mercantile Bancorporation, General American Life Insurance Co., LaBarge, Inc.

Kenneth T. Derr (48), Vice-Pres. of Co.; Pres. & Chief Exec. Off., Chevron U.S.A., Inc.

Lawrence W. Funkhouser (63), Vice-Pres., Explor. and Prod., of Co.

John R. Grey (62), Pres. of Co.; Dir., Bank of American NT&SA and BankAmerica Corp.

Kenneth E. Hill (69), consultant to Blyth Eastman Paine Webber, Inc.

Carla Anderson Hills (51), partner, law firm of Latham, Watkins & Hills; Dir., International Business Machines Corp., The Signal Companies, Inc., Corning Glass Works.

George M. Keller (61), Chmn. of Bd. & Chief Exec. Off. of Co.; Dir., First Interstate Bank of Calif., First Interstate Bancorp.

Charles W. Kitto (63), Vice-Pres., Logistics and Trading, of Co.

James E. Lee (63), Vice-Chmn, of Co.; Chmn., Pres. & Chief Exec. Off., Gulf Corp. & Gulf Oil Corp., Dir., Joy Manufacturing Co., Pittsburgh National Bank, PNC Financial Corp., Gulf Canada Ltd., the American Petroleum Institute and West Penn Hospital.

W. Jones McQuinn (61), Vice-Pres., Foreign, of Co.

Charles M. Pigott (55), Pres., Dir. and Chief Exec. Off., PACCAR Inc.; Dir,. The Boeing Co.

Charles B. Renfrew (56), Vice-Pres., Legal Affairs of Co.

George H. Weyerhaeuser (58), Pres. and Dir. Weyerhaeuser Co.; Dir., The Boeing Co., SAFECO Corp.

John A. Young (52), Pres., Dir. & Chief Exec. Off., Hewlett-Packard Co., Dir., Wells Fargo Bank and SRI International.

At the same annual meeting that approved the Chevron takeover of Gulf and so lent Chevron support to Marxist Angola, a Chevron director resigned. This was David Packard, Chairman of Hewlett-Packard and a former Secretary of Defense. There is no record that Packard protested either Gulf support of Marxism or objected that Chevron should not join the band of American corporations who have aided world revolution. We doubt that Packard resigned on grounds of principle, because Packard was an Overseer of the Hoover Institution and Chairman of its Financial Committee back in the early 1970s when Hoover Institution Director W. Glenn Campbell attempted to put pressure on this author to stop publication of the earlier version of this book, *National Suicide: Military Aid to the Soviet Union.*

Another interesting facet to this story of Soviet-multinational cooperation is in the amount of federal taxes paid by these giant firms. In 1976 Gulf had a federal tax rate of 2%. In 1984 Gulf claimed a tax refund, even while showing a profit of $313 million. Gulf 1984 tax rate was **minus** 8.3% (with a refund of $26 million). So in the same year that Gulf contributed most of Marxist Angola's foreign exchange and paid Angolan taxes, it demanded $26 million refund from U.S. taxpayers. If payment of taxes is a measure of patriotism, then Gulf Oil allegiance is more than clear.

Identification of the Deaf Mute Blindmen

The severity of our charges in the Chevron-Gulf case suggest that we be doubly careful in identification of the deaf mute blindmen. Not all directors of multinational corporations fit the description. Some in fact decidely do not. Not all politicians and bureaucrats fit the bill, although it is hard to find exceptions in Department of State.

Let's take a paradoxical example to demonstrate the need for care. In the Reagan Administration both Secretary of Defense Casper Weinberger and Secretary of State George Schultz are former officers of Bechtel Corporation, the multinational construction firm. However, Weinberger is one of the few in Washington to understand technological transfers. George Shultz on the other hand has a long-time unchanging record in favor of continuing transfers. Yet both have been top officials within Bechtel Corporation.

While Department of State has never produced a single opponent of transfers, the Department of Commerce has — Lawrence Brady. And although Weinberger is at DoD, the same department has produced several deaf mutes from among its top officials. The previously cited David Packard, Chairman of Hewlett-Packard, is a major sponsor of Congressman Ed Zschau, a vocal active supporter of more aid to Soviet military power.

Identification has to be handled on a case by case, individual basis.

What is to be Done?

The basic solution to Chevron-Gulf and other cases is political. It will take pressure from grass roots Americans and indeed from all those who love freedom to knock sense into their elected representatives and then into successive administrations and finally into the deaf mute blind-men.

There is, however, a startpoint which is demonstrated by the Chevron-Gulf case. Department of Defense has shown realistic concern over transfers of technology and the military end-use in Soviet world ambition. Why doesn't DoD start to consider Western technology to the Soviets in the award of DoD contracts?

In brief, penalize those firms working both sides of the street. If Chase Manhattan wants to finance Soviet contracts, all well and good, but it should not also expect a piece of the U.S. Defense pie at the same time. If General Electric wants to sell to the Soviets, OK, let G.E. go ahead — within the law — but not also simultaneously benefit from DoD contracts. If Chevron-Gulf want to work hand-in-glove with Soviet military ambitions, they should not also be able to bid on U.S. government contracts and claim federal tax refunds when the firm is making substantial profit. When awarding DoD contracts, preference should be given to those U.S. and foreign firms who show enough sense to walk away from Soviet deals.

CHAPTER XII

Tanks

"In the building of bridges toward peaceful engagement nothing, of course, should be done which would threaten our national security."

William Blackie, Chairman,
Caterpillar Tractor Company

Production facilities for tanks and armored cars combine features required in automobile and truck production with those required for locomotive and tractor production. Consequently, agricultural tractor, locomotive and automotive plants can be — and are — used to produce tanks, although mass production of tanks requires equipment changes and new machine installations.

Such civilian-to-military plant conversions for tank and armored-car production have been successfully undertaken in automobile and locomotive plants in many countries. In the United States the Ford Motor Company, Cadillac, and Chrysler have mass-produced tanks. In Italy the Fiat Company and in France the Renault Company, Citroen, and other automobile manufacturers have produced tanks. In England the Vauxhall Motor Company was a tank producer in World War II. Among locomotive manufacturers both Baldwin Locomotive and American Locomotive in the United States produced tanks in World War II. Tractor plants have been successfully converted to tank production — for example, Massey Harris in the United States and all Caterpillar tractor plants in the Soviet Union.

A tractor plant is well suited to tank and self-propelled gun production. The tractor plants at Stalingrad, Kharkov, and Chelyabinsk, erected with almost complete American assistance and equipment, and the Kirov plant in Leningrad, reconstructed by Ford, were used from the start to produce Soviet tanks, armored cars, and self-propelled guns. The enthusiasm with which this tank and armored-vehicle program was pursued, and the diversion of the best Russian engineers and material priorities to military purposes, have been responsible for at least part of the current Soviet problem of lagging tractor production and periodic famines.

Since 1931, up to a half of the productive capacity of these "tractor" plants has been used for tank and armored-car production.[34]

In both the State Department and the German Oberkommando files, there are reports confirming the **planned** adaptability of Soviet general-

[34]While, for example, the Soviet PT-76 tank used in 1972 in Vietnam came from the American-built Stalingrad plant (now called Volgograd), the exact percentage of the plant's capacity used for tank production is unknown.

equipment plants for war use — that is, the plants were originally planned for war use. For example, "The heavy industry plants are fitted with special attachments and equipment held in reserve which in a few hours will convert the plants into munitions factories."[35]

Tank assembly, like the production of automobiles, trucks, and tractors, is normally straight-line with components fed into the main assembly operation from subassembly lines. The components required for tanks are usually peculiar to such weapons. Tank power-plants and tank power-trains are not normally used in commercial-type vehicles in the West. However, in the Soviet Union commercial engines have been used in tank installations by combining two power plants in one tank, for example, the SU-76 self-propelled gun with two standard Dodge automobile engines, or by adapting aircraft engines.[36]

Machinery and machine tools for tank production are similar to those used in the production of heavy equipment, with additional special-purpose tools. Transfer machines required for automotive-type engines, large boring mills, large planers, radial drills, and heavy welding equipment are also utilized.

Consequently, any automobile, truck, locomotive or tractor production plant with straight-line assembly operations can be converted to the mass production of tanks by the addition of certain specialized equipment and by utilizing components and subassemblies made elsewhere for the specific tank vehicle to be assembled.

Soviet tractor plants were established in the early 1930s with major U.S. technical and equipment assistance. The Stalingrad tractor plant was completely built in the United States, shipped to Stalingrad, and then installed in prefabricated steel buildings also purchased in the United States. This unit, together with the Kharkov and Chelyabinsk plants and the rebuilt Kirov plant in Leningrad, comprised the Soviet tractor industry at that time, and a considerable part of the Soviet tank industry as well. During the war, equipment from Kharkov was evacuated and installed behind the Urals to form the Altai tractor plant, which opened in 1943.

Three postwar tractor plants were in operation by 1950: the Valdimir plant opened in 1944, the Lipetsk plant in 1947, the Minsk plant and the Kharkov assembly plant in 1950. This was the basic structure of the Soviet tractor industry in the 1960s and 1970s.

These plants produced tractors with a heavy emphasis on crawler (caterpillar-tread) models rather than the rubber-tired tractors more commonly used in the United States. The 1959 USDA technical delegation estimated that 50 percent of the current output was in

[35]Horace N. Filbert, "The Russian Industrialization Program" (unpublished manuscript in the Hoover Institution at Stanford University), p. 3.

[36]Aberdeen Proving Grounds Series, *Tank Data I* (Old Greenwich, Conn.: WE Inc., n.d.), p. 143.

crawler models, as contrasted to only 4 percent in the United States. The military implications of this product mix is obvious from Table 12-1.

Table 12-1
Soviet Tank Models Produced in Tractor Plants

Tank Model Number	Tractor-Tank Plant:	Where Tanks were used:
T-26 (8-ton) A, B, C versions	Ordzhonikidze (Kharkov)	Spanish Civil War Manchuria Finland
T-37 (3-ton)	Stalingrad Chelyabinsk	Russo-Finnish War World War II
T-32 (34-ton)	Kirov works (Leningrad)	Russo-Finnish War World War II
BT (12-ton)	Chelyabinsk	Spanish Civil War Russo-Finnish War
BT-28 (16-ton)	Chelyabinsk	Tusso-Finnish War
PT-76	Volograd (Stalingrad)	Indo-Pakistan War Vietnam

The Development of Soviet Tank Design

Before World War II Soviet tanks derived from American, British, and, to a lesser extent, French and Italian designs. Little German design influence can be traced in the period before 1939. During the 1920s and 1930s the Soviets acquired prototypes from all tank-producing countries and based development of Soviet tanks upon these foreign models. The Soviet tank stock in 1932 is shown in Table 12-2.

From this stock of Western models, together with technical-assistance agreements with foreign firms and the continuing purchase of foreign prototypes, the Soviets developed a formidable tank force for World War II.

The Carden-Lloyd (the predecessor of the British Brengun carrier of World War II) was a 1.69-ton machine-gun carrier first produced by Vickers-Armstrong, Ltd., in 1929. The Mark-VI model sold to the Soviets had a Ford Model-T 4-cylinder 22.5-horsepower water-cooled engine and a Ford planetary transmission. This became the Soviet T-27 light reconnaissance tank produced at the Bolshevik plant in Leningrad.

The Ordzhonikidze Tractor Plant at Kharkov started work on the T-26, based on the British Vickers-Armstrong 6-tonner at about the same time. There were three versions — A, B and C — of which B and C became the standard Soviet models produced until 1941. Similarly, the Soviet T-37 and T-38 amphibious vehicles were based on the Carden-Lloyd Amphibian, known as the Model-A4 E 11 in the British Army.

Table 12-2
Soviet Tank Stock in 1932

20 Carden-Lloyd Mark-VI	Vickers-Armstrong Ltd. (U.K.)
1 Fist Type-3000	Fiat (Italy)
20 Renault	Renault (France)
16 "Russian-Renaults"	Made in France, modified in USSR
70 light tanks	Vickers 6-ton, Alternate-A (U.K.)
40 Vickers Mark-II	Vickers-Armstrong Ltd. (U.K.)
2 Christie M-1931	U.S. Wheel Track Layer Corp. (U.S.A.)
8 Medium Mark-A	Vickers-Armstrong Ltd. (U.K.)

Sources: R.E. Jones et al., *The Fighting Tanks Since 1916* (Washington, DC: National Service Publishing Co., 1933), p. 173.

R.M. Ogorkiewicz, "Soviet Tanks," in B.H. Liddell Hart, ed., *The Red Army* (New York: Harcourt, Brace and Co., 1956).

Walter Christie, a well-known American inventor with numerous automotive and tank inventions to his credit, developed the Christie tank — the basis of World War II American tanks. Numerous Soviet versions of Christie tanks and armored vehicles were produced in the late 1920s and 1930s. Two chassis of the Christie M-1931 model medium tank (MB) were purchased by the Soviet Union in 1932 from the U.S. Wheel Track Layer Corporation. After further development work this became not only the Soviet T-32 (the basic Soviet tank of World War II), but also several other development models in the USSR: first the BT (12 tons), followed by the BT-5 and the BT-28 produced at Chelyabinsk.

The Soviet T-34 and the American M-3, both based on the Christie, had the same 12-cylinder aero engine, a V-type Liberty of 338 horsepower. Ogorkiewicz comments on the Christie model series as follows:

> The power-weight ratio was actually higher than could be efficiently used, but the Russians copied it all and confined their development largely to armament, which increased from a 37-millimeter gun on the original models of 1931, to 45-millimeter guns on BT-5 of 1935, and eventually to short 76.2-millimeter guns on some of the final models of the series.

Both the Soviet T-28 medium 29-ton tank and the T-35 45-ton heavy tank resembled British models — the A-6 medium tank and the A-1 Vickers Independent, respectively. Imported French Renault designs also contributed to Russian tank knowledge. During the 1933 entente between France and the Soviet Union, the Renault Company delivered $11 million worth of "small fast tanks and artillery tractors" to the Soviet Union and supplied experts from the Schneider works and Panhard Levasseur, skilled in the armored-car and tank field.

The Famous T-34 Medium Tank

The Soviet T-34 and the modified T-34/85 were first introduced in World War II and used extensively against Americans in the Korean War. The model was later used in the Hungarian revolt and in "wars of liberation" to the present day. The T-34 is an excellent design and a formidable weapon. It emphasizes the ability of the Soviets to design weapons while still dependent on the West for production facilities and basic technical advances.

In 1931 the Russians bought two Christie tanks from the U.S. Wheel Track Layer Corporation in the United States. The Russians copied these, built Christie tanks, and then incorporated the Christie suspension system into the T-34. The first Russian Christies had the same engines as the U.S. Christie — a Liberty 12-cylinder V-type of 338 horsepower with forced-water cooling. In the 1920s the Chase National Bank of New York (now Chase Manhattan) attempted to arrange illegal export of large quantities of these Liberty engines to the Soviet Union at the price of $2,000 each.

In any event, the T-34 incorporated the Christie suspension from the United States, but generally used a 500-horsepower V-type diesel developed from the German B.M.W. diesel engine. The T-34/85 was the T-34 with significantly increased firepower. Ball bearings on the T-34 and T-34/85 were manufactured on Swedish equipment.

The welding work on the T-34 was at first immensely crude, but as *The Welding Engineer* (Dec. 1952) pointed out: "The T-34 was designed with one idea in mind — to provide firepower. Any humanitarian considerations, like protection of the crew, are purely secondary."

The original T-34s were built from several million tons of armor-plate imported from the United States. In July 1934 Henry Disston & Sons, Inc. requested War Department permission to accede to a Soviet request, "in training their technicians to make tank armorplate of the same quality as they now make for this Government" (Russia 400.114, War Office).

The T-34 was followed by the improved T-44 and then by the T-54, with the basic T-44 chassis and using Christie-system torsion-bar suspension. This was the standard Soviet tank until recently; it was used in the Hungarian revolt in 1956, in South Vietnam in 1972, and still in use in most Warsaw Pact countries.

The Red Army has always used diesel engines in its medium and heavy tanks. This tank-engine series is the V-2 and V-12 water-cooled, rated to 550 brake horsepower (bhp) at 2,150 revolutions per minute (rpm). According to Ogorkiewicz, the original Russian water-cooled V-12 engine was a successful diesel adaptation of contemporary aero-engine designs. Used on all Soviet medium and heavy tanks up to World War II, it was a large 2,860 cubic-inch engine, based on a German B.M.W. aircraft design, and developed about 500 bhp. Soviet emphasis on diesels has continued since World War II, while other Soviet armored vehicles have used automobile gasoline engines. The T-70 light tank uses two GAZ-202 70-horsepower engines from the Ford-Gorki plant. The SU-76 self-propelled gun also used two engines of the same Ford type geared together.

The Soviets have continuously developed the same engine for their battle line tanks including the T-34, T-44, T-54, T-62 and T-72. These engines originated with a Hispano Suisa engine originally developed in the 1930s as an aircraft engine. The latest T-62s are manufactured in three gigantic plants at Nizhny Tagil, Omsk and Kharkov. It has a power displacement of 9.93 kw/liter, a mean piston speed at maximum rpm of 12.6 m/s for the left bank and 13.1 m/s for the right bank. The BMEP at maximum torque is 698 kPz. The weight to power ratio is 2.6 kg/kw. The weight/displacement is 25.8 Kg per liter and the specific fuel consumption is 210 g/kwh.

The Soviets prefer not to use imported parts for field military end uses. However, they do reverse engineer the latest advances in Western military technology obtained by illegal purchase or espionage. Examples are the gunner's telescope on the T-62 which uses the same multi roundscales and movable index line found on the M-46s (U.S.) T-152 co-axial telescope. The T-64 and the T-72 has an analogue ballistic computer and a coincidence rangefinder comparable to those found on U.S. M-48 and M-60 tanks. These were probably obtained from U.S. equipment captured in Vietnam.

DMB Pleas of Ignorance

A common statement from the deaf mute blindmen is that their contracts with the Soviet Union have no military potential and cannot therefore do any harm to our national security. These statements are not only false, but these multinational businessmen know they are false.

We cited William Blackie, Chairman of Caterpillar Tractor Company, in the epigraph to this chapter to the effect that "building of bridges" should not "threaten our national security." Yet Mr. Blackie knows full well the use to which the Soviets put the Caterpillar Tractor plants in the Soviet Union and the military use of Caterpillar tractor reverse engineered from U.S. models. His statement is bland subterfuge.

The American-built tractor plants at Stalingrad, Kharkov, and Chelyabinsk were the **only** major tank producers of the Soviet Union up to World War II. Tank output for 1938 (a year for which complete figures are available) is shown in Table 12-3, with the percentage tank output on early American-built tractor plant.

Table 12-3
Overall Annual Production of Soviet Tanks in
U.S.-Built Tractor Plants (1938)

Percentage of Total Produced in Each Plant	Construction of Plant by:	Origin of Tank Model
Chelyabinsk: 28.9 percent	U.S. firms (1933)	Christie (U.S.A.) Carden-Lloyd (U.K.) Vickers (U.K.)
Stalingrad: 32.8 percent	U.S. firms (1930)	Carden-Lloyd Vickers-Armstrong
Kharkov: 38.3 percent	U.S. firms (1931)	Vickers-Armstrong

The U.S.-Built Stalingrad "Tractor" Plant

In March 1929 a delegation of thirteen Soviet engineers arrived in the United States and in cooperation with several American companies outlined a plan for a plant to produce 50,000 Caterpillar-type tractors a year. "The entire designing of the Stalingrad . . . tractor plant . . . was carried out in the United States. . . . While preliminary work on the site of the Stalingrad Tractor Plant had been conducted for some time, the actual work on the construction of the principal departments started only in June when the plans arrived from the United States.[37]

The Stalingrad Tractor Plant, the largest in Europe, was a packaged factory built in the United States, dismantled, shipped to the USSR, and re-erected at Stalingrad under the supervision of American engineers. All its equipment was manufactured in the United States by some eight firms; it went into production with the Harvester 15/30 model and the T-37 3-ton tank.

The Stalingrad Tractor Plant was the first of three massive plants for the production of tractors in peace and tanks in war. It was built in every sense of the word in the United States and was reassembled in Stal-

[37]Amtorg, *Economic Review of the Soviet Union,* 5:7 (Apr. 1, 1930), 135.

ingrad by 570 Americans and 50 Germans. The plant was delivered in component parts, installed in a building supplied by McClintock & Marshall, and erected under the supervision of John Calder of the Austin Company.

Za Industrializatsiiu pointed out that "it is very important to note that the work of the American specialists . . . was not that of consulting but of actually superintending the entire construction and the various operations involved."[38]

Each item of construction and equipment was the responsibility of a major U.S. firm: the design of plant was by Albert Kahn, Inc.; the design of the forge shop was by R. Smith, Inc.; the design of the foundry was by Frank C. Chase, Inc. Equipment for the cold-stamping department came from Niagara and Bliss; equipment for the heat-treating shops was by Rockwell; equipment for the power station by Seper and Westinghouse.

Equipment for chain-belting in the conveyor system was by Chain Belt Co., and the supply of buildings by McClintock & Marshall.

The Stalingrad Tractor Plant, therefore, was American in concept, design, construction, equipment, and operation. It could just as easily have been located outside Chicago, producing Harvester tractors, except for the placards claiming "socialist progress" and its massive tank quota.

It is worthwhile to recall that the contemporary Soviet press was quite open about this U.S. assistance. For example, an article in *Za Industrializatsiiu* drew three conclusions: first, that the preparation of the plans for the Stalingrad plant by American engineers with "participation" by Soviet engineers made completion of the plant possible within a "very short time"; second, that work and training by Soviet engineers in the United States resulted in a "considerable improvement in engineering processes" and the application of American standards; and third, that work in the United States gave the Soviets a firsthand opportunity to study American tractor plants and verify data on the operation of American machine tools.

As early as 1931 the Chain Belt Company representative, who was installing a conveyor system at Stalingrad, reported that the newly opened tractor plant was making "small tanks." In 1932 A.A. Wishnewsky, an American whose work took him into many Soviet factories, reported that the principal emphasis in all these new tractor plants was on the production of munitions and military supplies. In all factories, he stated, at least one department was closed, and he would from time to time run across "parts and materials for military production." This was particularly true of Tractorostroy (**sic**), where emphasis

[38]July 5, 1930.

was placed on the production of tanks rather than tractors. "In his opinion, at least for the time being, the development of tractor production there has been designed to lead up the production of tanks for military purposes."

Such early reports were confirmed a few years later by German intelligence, which reported that in 1937-38 the Stalingrad Tractor Plant was producing a small 3-ton armored car, a self-propelled gun, and the T-37 tank, which was patterned on the British A-4 Ell.

Recently, Stalingrad (now Volgograd) has produced the PT-76, an amphibious unit in the Soviet tank stock, which was used in South Vietnam and elsewhere around the world.

The U.S.-Built Kharkov "Tractor" Plant

The Kharkov "tractor" plant was identical to the Stalingrad plant. The original intention was to build Kharkov as an all-Soviet undertaking, but American engineers were called in at a very early point. Leon A. Swajian, a well-known engineer in the United States, became chief construction engineer and was subsequently awarded the Order of Lenin for his work. Swajian commented that no other construction job in his experience had required so much work in a single year, and that in the United States such giant plants are not built all at once, but a few departments at a time, by subcontract. The same American supervising engineers and similar U.S. construction equipment and methods were used. Thus, Swajian explained:

> Ford's River Rouge plant was more than a dozen years in building. When I took charge [at River Rouge] it was already partly built; I worked there six or seven years and when I left construction was still in progress. But in the U.S.S.R. with government financing and no other plants from which to buy spare parts, with the plant dependent on itself — down to the smallest operation on the basic raw material — the whole plant must be built at once. And very swiftly too, if it is not to tie up capital too long. The Kharkov job was pushed to completion more swiftly than any job I have ever had to do with.[39]

As at the Stalingrad and Chelyabinsk tractor plants, the equipment was almost all foreign — either American or German, patterned after American makes. No equipment at the Kharkov plant was Soviet-made. The forge shop contained $403,000 worth of American forging machines and dies, and the heat-treating equipment and automatic furnace-temperature controls were supplied by Leeds and Northrup of Philadelphia. A report in late 1932 from the Kharkov Tractor Plant by Ingram D. Calhoun, an engineer for the Oilgear Company of

[39]Amtorg, *op. cit.*, 6:18 (Sept. 15, 1931), p. 413.

Milwaukee, stated that Kharkov was turning out eight to ten tanks a day and tanks took precedence over tractor production. Operators were being trained "night and day . . . they can fool the tourists but not the foreign engineers," Calhoun added.[40]

By 1938 Kharkov was producing self-propelled guns, armored cars, and the T-26 tank, which was patterned after the British Vickers-Armstrong 6-tonner.

The U.S.-Built Chelyabinsk "Tractor" Plant

The Chelyabinsk "tractor" plant was started in 1930, without foreign technical assistance, as another duplicate of the Stalingrad Tractor Plant. One year later, in March 1931, a letter to the Soviet press, signed by thirty-five Chelyabinsk Tractor Plant engineers and economists, charged that the project was "on the verge of total collapse."

American engineers, including John Calder, the expert troubleshooter, were then called in to take over construction of the plant and initial operating responsibility. A pilot plant was established and operated by John Thane and an American assistant, both of whom were former employees of the Caterpillar Company. The chief consulting engineer from 1931 to 1933 was Edward J. Terry. The Stalinets S-60 tractor produced was an exact copy of the Caterpillar 1925-31 model. Ex-Caterpillar engineers supervised operations. In May 1933 practically all the machine tools and production equipment in the plant were American, British, or German.

By 1937 the plant employed about 25,000 workers. The only tractor produced between 1933 and 1937 was the Stalinets (Caterpillar) S-60, a 50-horsepower (drawbar) model of the crawler type. About 6,460 were produced in 1937, a long way from the planned 50,000 per year. In 1937 the production model was changed to the Stalinets S-65, which was a Caterpillar-60 with slightly increased horsepower and a diesel engine. A total of just over 3,000 were produced, including another model with a gas generator.

The Chelyabinsk Tractor Plant was also producing tanks of the BT series, which was patterned on the American Christie. Monthly output in 1938 consisted of thirty-two of the 12-tonners and 100 of the BT-38, a 16-tonner.

Thus, not only were all three of the new American-built tractor plants producing tanks throughout the 1930s, but they were by far the most important industrial units producing this type of weapon. Today, these plants still can, and do, produce tanks. Yet multinational businessmen continue to blandly assert that their dealings have no impact on our national security.

[40]U.S. State Dept. Decimal File, 861-5017 — Living Conditions/576, Dec. 28, 1932.

CHAPTER XIII

Why the DMBs Aid Soviet Ambitions

"One of the reasons that Americans are often confused about the Soviet Union is that Soviet manipulation of diplomats, who senselessly limit their contacts, and Sovietologists, who depend on the Soviet authorities for their visas, has given rise to a conventional wisdom about the 'reasonableness' of the Soviet Union that bears no relationship to reality."

(David Satter, *Wall Street Journal*,
October 22, 1985)

The deaf mute blindmen have an outstanding quality — an ability to have policy makers listen to and implement an irrational policy based on fantasy and zero evidence, with no relationship to empirical reality.

The public plea of internationlist businessmen is that trade brings peace. This is usually phrased along the following lines. This example is from Maurice Stans, friend of Richard Nixon, former Secretary of Commerce, and Wall Street investment banker: "History has shown that where there is increasing trade between countries . . . there is a tendency toward increasing understanding."

Unfortunately for Maurice Stans' credibility, no historian or politician has ever produced **evidence** that trade necessarily and automatically leads to increased understanding. The statement has no empirical justification at all. It sounds good, that's all.

Why not? Because no such evidence exists. While it is true that peace encourages trade, this does not mean that trade encourages peace. In fact, trade has often enabled aggressive countries to go to war — witness the aviation gasoline and steel scrap shipments to Japan and Standard Oil's agreements on hydrogenation patents with I.G. Farben of Germany prior to World War II.

The blunt truth is that trade with the Soviet Union from 1917 to the present has built the Free World an enemy of the first order. Moreover, the technological component of this continuing trade enables the Soviet Union to pursue its programs of world conquest. It costs the American taxpayer $300 billion a year to counter this Soviet threat.

Obviously, claims that increasing trade is accompanied by increasing understanding are false. A more accurate characterization in the Soviet case appears to be that increasing trade is accompanied by increasing conflict (i.e., misunderstanding), because the Kremlin is encouraged to believe that capitalists are behaving as Lenin predicted they would behave. The concept of ."peaceful trade" with Communist countries assumes some positive causal link between trade and peace. It is a false assumption.

The overlooked link is the relationship between trade and casualties, and is contained in one word: **technology.**

From the Soviet viewpoint, the benefit of U.S.-Soviet trade is the acquisition from the United States of the technology required for building and maintaining the Soviet power base. This subsidized technology is the basis of the Soviet military structure, as is proved by the Soviets' own statements. The armaments produced by Russia's American-subsidized military-industrial complex are used to promote Soviet global expansion.

There are four logical steps in this process:[41]

1. Trade between the United States and the USSR (with its key technological component).
2. Consequent buildup of the Soviet military-industrial complex.
3. Use of the U.S.-subsidized Soviet military-industrial complex to provide inputs for Soviet armaments production.
4. Use of these armaments against the United States and its allies.

This is the "trade-technology-armaments-war" cycle. It suggests that "peaceful trade" promoted by successive administrations should be entitled "war trade."

If any of the links in this cycle can be broken (or shown to be false), the argument collapses. However, no link is demonstrably false. Indeed, there is evidence that the trade-casualties link is stronger than even the most pessimistic have envisaged.

We trade freely with European nations because these nations have no deceitful intent to use the technological component of this trade against the United States. Nor has the United States any hostile intent towards these European countries. Free trade is vital and necessary, and both sides benefit from its advantages. However, gains from free trade utlimately depend on **intent.** Where countries are potentially hostile and make hostile use of imported technology, the gains from free trade must be modified by the hostile intent of the trading partner. In other words, "peaceful trade" is only peaceful if Soviet intent is peaceful.

Why isn't this information and logic made available to the public? Then at least the DMB would have official opposition to their spurious arguments. For some reason or another the Washington bureaucracy in State and Commerce has by and large, with some exceptions, taken a position inconsistent with its own files.

---

[41]Steps 1 and 2 have been fully demonstrated in the author's three-volume study, *Western Technology and Soviet Economic Development.* See Bibliography.

The Bureaucrats' View of "Peaceful Trade"

The State and Commerce Departments have consistently rejected the argument that "peaceful trade" can assist Soviet military objectives. For example, a 1969 State Department leaflet asserts (under the heading "US exports do not help Hanoi"):

> Over two-thirds of our exports to the Soviet Union and East Europe are foodstuffs and raw materials for consumption within their economies. There is no evidence that our exports of such goods to these countries release resources for manufacturing war materials for North Vietnam.[42]

What is wrong with this statement?

The Soviet Union needs — and receives — U.S. technology, not "foodstuffs and raw materials." The bureaucracy may claim U.S. wheat does not go to Vietnam and Cuba, but it avoids the crucial point that export of our wheat to the Soviets released Soviet wheat for export to Hanoi. No economist will deny that our technical transfers release Soviet domestic resources for armaments production. The State Department assertion is therefore a compound of distortion and ignorance.

These unsubstantiated distortions are given to Congress as verifiable truth. For example, Edwin M. Martin, Assistant Secretary of State for Economic Affairs, made the following statement to Congress in 1961: "I don't think there is convincing evidence that the net advantage to the Soviet Union of the continuation of trade is a major factor — or a particularly significant factor in the rate of their overall economic development in the long term."[43] This is an untrue statement.

There is also confusion concerning the Soviet practice of copying. For example, the following exchange took place before a congressional committee in 1961.

> **Mr. Lipscomb:** Does the Department of Commerce feel that Russia has developed a great deal of their agricultural equipment from prototypes obtained both legally and illegally from the United States?

> **Mr. Behrman:** No sir, I don't think that the evidence we have indicates that the equipment that they themselves produce copies — that they produce copies of equipment which we have supplied.[44]

Again, a wildly untrue statement.

[42]U.S. Dept. of State, Public Information Series P-310-369 (Washington, DC, 1969).

[43]Edwin M. Martin, Assistant Secretary of State for Economic Affairs, before the House Select Committee on Export Control, Dec. 8, 1961.

[44]U.S. House of Representatives, Select Committee on Export Control, Investigation and Study of the Administration, Operation and Enforcement of the Export Control Act of 1949, and Related Acts (H.R. 403), 87th Cong., 1st sess., Oct. 25, 26, and 30, and Dec. 5, 6, 7 and 8, 1961, p. 403.

Even well-informed members of Congress have taken positions directly opposed to the evidence. Senator Jacob Javits of New York commented: "Trade with the West as a general matter must necessarily be a marginal factor in the performance and potentialities of the Soviet economy."[45]

There exist, presumably unknown to Senator Javits and the others quoted above, three volumes of detailed evidence that totally refutes these statements.[46]

A popular book in the 1930s was *You Can't Do Business with Hitler!* The moral and national-security arguments in this book apparently apply only to Hitler's brand of totalitarianism. There is extraordinary inconsistency in the treatment of Hitlerian totalitarianism and Soviet totalitarianism. Indeed, there is some direct evidence and a great deal of indirect evidence that the policymakers in Washington do not view the Soviet Union as a totalitarian power at all.

At the end of World War II, the conclusion of an interagency committee on German industry, with members from the State and Commerce Departments, was as follows:

> The Committee is unanimously of the opinion that the major force for war represented in a motor vehicle industry is its availability as a major machine shop aggregation, under able management and engineering, which can be turned, by conversion, to production of an extremely wide variety of military products. Its role as a producer of combat and military transport vehicles ranks second in significance.[47]

Yet **today** State and Commerce argue that the export of equipment for a motor vehicle industry is "peaceful trade" — **even when military vehicles produced by previously exported technology are photographed in Vietnam, Cuba, Afghanistan, Angola and Nicaragua.** Obviously, no amount of hard evidence can shake these people from their illusions. The policymakers are locked onto a brand of totalitarianism which, to them, is morally and strategically acceptable.

U.S. assistance to the Marxist brand of totalitarianism is not limited to the Soviet Union. In 1971, for example, the State Department attempted to help Allende, the Marxist president of Chile, to purchase aircraft and paratroop equipment in the United States (*Indianapolis News*, July 1971). Three months later Allende attempted to impose complete Marxist control in Chile. Once again the State Department wants to help a Marxist clique to impose its rule on an unwilling population.

[45]*Congressional Record*, Senate, Vol. 122, pt. 9 (89th Congress, 2nd session), May 24, 1966, p. 11233.

[46]Antony C. Sutton, *Western Technology and Soviet Economic Development* (See Bibliography).

[47]Foreign Economic Administration. *Study by Interagency Committee on the Treatment of the German Automotive Industry from the Standpoint of International Security* (Washington, DC, 1945).

The assertion that we should exchange our technology for Soviet raw materials will not stand penetrating analysis. Soviet raw materials only become competitive to our enormous low-grade domestic reserves if their extraction is subsidized by U.S. loans and guarantees — to the disadvantage of the American taxpayer. From the national security viewpoint, the "exchange" is absurd. Once our technology is passed to the Soviet Union it cannot be reclaimed, it becomes an integral part of their military industrial operations. But Russian raw material supplies, developed with our assistance, can be cut off any time the Soviets wish. In other words, once again our policymakers exchange something for nothing and charge it to the American taxpayer and citizen.

Useless Pinpricks as Policy

In early May 1985, President Reagan embargoed U.S. trade with Nicaragua. This followed what a State Department official described as a "disturbing pattern of behavior" on the part of Nicaragua. But the White House missed a point — and an opportunity.

Nicaragua and its "disturbing behavior" is a mere incident in a vast ever-expanding global game. Moreover, the Nicaraguan embargo has loopholes that make it virtually ineffective. For example, foreign affiliates of U.S. firms are not embargoed. In brief, if the invoice is on foreign subsidiary, that's legal. And don't think American firms won't scramble for the loophole.

The **global** problem is not pinprick Nicaragua — it is the unholy alliance of the multinationals and the Soviet Union.

Follow these logical steps:

- The Sandanista revolution only survives because of Marxist suppression and Soviet logistical backing.
- The Soviets survive and have become militarily awesome because the West has a sixty-year-old policy of technical, economic, financial and even political subsidy of the Soviet Union.
- The multinationals are the mainspring for technical and economic subsidy, from Kama River to the latest CDC computers. And they are more than willing to bend the truth to make a buck.
- Our political establishment, as exemplified in the Council on Foreign Relations, is the mainspring behind political subsidy of the Soviet Union, as exemplified in the Yalta agreement, National Security Memorandum No. 68 and so-called "detente."

Under this system the Soviet Union is never held fully accountable for its international acts. For example:

- The Soviets illegally occupy the long independent countries of Latvia, Lithuania and Estonia.
- The Soviet Union has occupied the long independent nation of Afghanistan.
- Soviet troops are based in Poland, Hungary, Rumania and Czechoslovakia.
- Soviet troops are active in Cuba, Angola, and a dozen other hot spots around the world including Nicaragua.

This adds up to global aggression. Yet the political establishment and its Congressional allies told us that peaceful coexistence and detente with the Soviets would bring peace.

The American public has never been told by any Administration the Soviet definition of "peaceful coexistence." That detente in the Soviet interpretation includes the RIGHT to continue so-called wars of liberation and to further encroach upon the Western world. Western governments have persisted in the illusion of detente as peace, knowing full well that the Marxist interpretation of peaceful coexistence is aggression, that peace is indeed war.

It is this official inaction that has encouraged the deaf mute blindmen to further Soviet ambitions for their own personal gain.

Multinational Businessmen and the Politics of Greed

Who benefits from technological transfers to the Soviet Union? Basically two groups: the Soviets themselves and the comparative handful of companies, almost all multinationals, who set up and profit from Soviet contracts.

Soviet business is profitable. Some American companies have been sucking in this profit since the Bolshevik Revolution of 1917. Chase National was involved in illegal exports to the Soviet Union in the early 1920s, while Chase Manhattan was the lead financing bank for the Kama River project. Caterpillar Tractor was involved in the First Five Year Plan in 1930, and still involved in the Siberian pipeline "Russia No. 6" project in the 1980s.

The hundreds of examples cited here are not accidents. Many individual congressmen and persons outside government have attempted to stop the export of military goods to the Soviet Union in the face of business pressures. Samuel Gompers tried in the 1920s. U.S. Navy officers risked court-martial in the 1930s in their effort to stop President Franklin D. Roosevelt from approving the sale of military equipment to the Soviets. The Senate investigated the "machine tools" case in 1945-46 when Commerce officials tried to send tank-gun milling equipment to the Soviet Union. Warnings were raised in the 1950s by Congressman Lipscomb and others about the results of building up Soviet military strength. In the 1960s individual American firms and engineers protested against the export of ball bearings processing equipment which could **only** be used to process bearings for Soviet missile-guidance systems. In the 1980s Department of Defense has raised a clear warning.

Are we to believe that Mr. Nicholaas Leyds, general manager of Bryant Chucking Grinder Company of Springfield, Vermont, did not **know** that his grinding machines are used to process the races for ball bearings in missile-guidance systems? When Bryant has sold the same machine to the U.S. government for the same purpose?

Or that the Hammer family does not know the results of its seventy year association with the USSR? In 1919 Dr. Julius Hammer was a member of the Executive Committee of the Communist party of the United States and in 1985 Armand Hammer (his son and early business partner) is described as America's "No. 1 Capitalist."

It is not an accident. The deaf mute blindmen have always known the exact end-uses of the exported equipment. They have been brazen enough to persist in building up the Soviet military-industrial complex while pleading "civilizing the Bolsheviks" (1918) and "peaceful trade" (1985).

Above all there are several disturbing aspects to this whole business of selling our technology to the Soviets. First, multinational banks and business groups have their own people in all administrations pushing for their self-interested profitable projects. Second, the more prominent of these businessmen (Armand Hammer and David Rockefeller come to mind) deal directly with the Soviet Union almost as independent fiefdoms.

The business and personal relationships of Henry Kissinger and the Rockefeller family, for instance, are too close for administrative objectivity. Neither has Henry Kissinger been on the receiving end of a bullet, he doesn't have to pay the price for his policies. **Indeed, in most cases policymakers have never undertaken any occupation where the cost of a wrong decision fell upon their own shoulders — and that may be one of the major problems.**

The following extract is from the conclusion of Kissinger's book *Nuclear Weapons and Foreign Policy* (p. 431):

> A statesman must act as if his inspirations were already experience, as if his aspirations were truth. He must bridge the gap between society's experience and his vision, between its tradition and its future.

Note the words "act as if." Act "as if" fantasy were reality is the clear implication. Act "as if" objectives (whatever they may be) are always moral.

In other words, anything is real if you wish it to be, and anything is moral if you want it to be. Is it then accidental that Kissinger's devastating detente policy was only beneficial to his proteges? (For example, the Rockefeller controlled Chase Manhattan Bank).

Then the amoral two-facedness of many DMBs has been commented on by several observers:

> At times it seemed that the business world preferred to ignore the policy makers in Washington. One prominent member of the council [The U.S.-U.S.S.R. Trade and Economic Council] remembers meeting after Nixon's resignation with four other industrialists, several admirals, and a member of the Joint Chiefs of Staff late in 1974. One admiral bewailed the military disadvantages U.S. forces suffered because of certain technology transfers. The next day, the same industrialist attended a private dinner in New York given by a major investment banking house. The guest of honor was Dzhermen Gvishiani, the Soviet official whose specialty was buying high technology in the West. One after another, the thirty other businessmen at the dinner rose to boast about the last big transaction they had made with Gvishiani. 'The dichotomy between the luncheon in Washington and the dinner in New York was just astounding,' the businessman recalled. 'These guys were bragging as if they had just sold a ton of frankfurters.'[48]

This atmosphere of "bragging as if they had just sold a ton of frankfurters" has been officially encouraged because until very recently, no official in any Administration has stood up in public and castigated the deaf mute blindmen. In recent years Secretary of Defense Casper Weinberger and Assistant Secretary of Defense Richard Perle have to some extent changed the nature of the public debate — but State and Commerce still drag their feet.

Motivated by the lure of profit and in the absence of official chastisement, the deaf mute blindmen have run the show where trade with the Soviets is concerned.

Through the Export-Import Bank they have received official, less than market loans, subsidised by the taxpayer, for projects with military potential. Ex-Im provided $153 million for Kama River, $180 million for Occidental's string of chemical plants — at half the prime rate. When this latter venture was approved, William J. Casey was Chairman of Export Import Bank even while publicly today Casey disclaims knowledge of our build-up of the USSR.

Other businessmen have appeared before Congress and made statements inconsistent with the bulk of the evidence. For example, we

[48]Joseph Finder, *op. cit.*

know that the Soviets have successfully reverse-engineered thousands of U.S. processes. Yet one businessman appeared before Congress and denied the usefulness of the process:

> **Senator Nunn:** You say reversing engineering is very difficult and time consuming even if they have the equipment itself, and all the plans and specifications?
>
> **Mr. Bell:** To put it in production. That is true. You can get a lot of valuable information off of it. But to put that equipment into production, even if you had what I said, they do not have the components. That is the key to the whole thing.
>
> The best system engineer in the world, that is what we do in system engineering, is only as good as the elements he has to work with. They do not have the components in the Soviet Union.[49]

Again Charles Lecht, former President of Advanced Computer Techniques Corp., tried to advance the line that the only Soviet interest was curiosity rather than technological need:

> Primarily I have come to the conclusion that the concept that the Soviets want U.S. technology because they can't make it themselves or they can't buy it themselves is a fallacious one. I have concluded that the Soviet technology establishment — in the country which produces more scientific literature than any other country in the world — is capable of producing what it wants. I have thus concluded that the Soviets have been looking for U.S. technology primarily because they want to find out how our military materiel work, our missiles, our aircraft, our radar, our sonar, and the like.[50]

In brief, the bottom line is greed and the greed has been allowed to reign supreme because official Washington has openly encouraged technological transfers.

[49]United States Senate, *Transfer of United States High Technology to the Soviet Union and Soviet Bloc Nations*, Hearings before the Permanent Subcommittee on Investigations, 97th Congress, Second Session, May 1982, Washington, DC, p. 49.

[50]Ibid., p. 230.

Treason

"Technology cannot be kept from the Soviets by endless paper-work requirements. Americans who ship crucial weapons and documents to them are criminals: criminals are glad to sign papers."

George Gilder, New York Times

Are these businessmen, the deaf mute blindmen internationalists, also guilty of treason? It is interesting to initially look at this question from the viewpoint of the other side, the Soviet side. Avraham Shifrin, a former Soviet Defense official, has a blunt conclusion. Shifrin calls the transfers treason and "they (the businessmen) should be shot" (page 21).

To take another example from the other side, the Marxist rebels in El Salvador, the Farabundo Marti National Liberation Front, receive aid from the Soviet Union. At the same time they claim that our aid to the elected Salvadoran Government makes our military advisors in El Salvador legitimate targets for assassination, i.e., the U.S. is an enemy just by virtue of economic subsidy. In brief, the other side interprets aid, even with no technological component, as equivalent to treason.

In the original version of this book, *National Suicide: Military Aid to the Soviet Union*, published in 1974, we declined to term subsidy to the Soviets as treason because the vital element of **intent** was missing. This conclusion was phrased as follows:

Do the actions described in this book [i.e. *National Suicide*] constitute "adhering" to these enemies, "giving them Aid and Comfort"?

The actions do not legally constitute treason. The Constitution defines the term strictly, for the intention of the framers, with good reason, was to deny Congress the right to interpret treason too freely. Moreover, the body of relevant case law is not substantial. The Cramer and Haupt cases after World War I suggest that both intent to commit treason and overt treasonable acts are required, in addition to thorough proof. While the actions described here could be interpreted as giving immediate "Aid and Comfort" to the Soviet Union, there is no specific evidence of intent, and intent is a vital requirement. Idiocy, inefficiency, intellectual myopia, and so on, do not suggest intent (p. 240).

We need to pose the question again, given the accumulating evidence of the last 11 years. Does this sequence of events and actions fall within the meaning of treason? Specifically, does military aid to the Soviet Union constitute "adhering to their enemies, giving them Aid and Comfort"? as defined in Article III, Section 3 of the Constitution of the United States.

Are the Soviets Enemies?

The Soviets have always been explicit about their intentions — so was Hitler in *Mein Kampf*.

Objective truth has no place in Communist morality, by their own statements. Any statement that will advance the cause of world communism is regarded as truthful, acceptable, and perfectly normal. As far back as 1919, Zinoviev put it well in a statement that applies to the Viet Cong and the Sandanistas as much as to the revolutionary Bolsheviks:

> We are willing to sign an unfavorable peace. It would only mean we should put no trust whatever in the piece of paper we should sign. We should use the breathing space so obtained in order to gather our strength.[52]

This immoral dogma — moral only in Marxist ideology — was emphasized by Joseph Stalin:

> Words must have no relations to actions — otherwise what kind of diplomacy is it? Words are one thing, actions another. Good words are a mask for concealment of bad deeds. Sincere diplomacy is no more possible than dry water or wooden iron.[53]

In 1955 the staff of the U.S. Senate Committee on the Judiciary examined the Soviet historical record and, not unexpectedly in the light of the foregoing statements, came to the following conclusion:

> The staff studied **nearly a thousand treaties** and agreements . . . both bilateral and multilateral, which the Soviets have entered into not only with the United States, but with countries all over the world. The staff found that in the 38 short years since the Soviet Union came into existence, its Government had broken its word to virtually every country to which it ever gave a signed promise. It signed treaties of nonaggression with neighboring states and then absorbed those states. It signed promises to refrain from revolutionary activity inside the countries with which it sought "friendship" and then cynically broke those promises. It was violating the first agreement it ever signed with the United States at the very moment the Soviet envoy, Litvinov, was putting his signature to that agreement, and it is still violating the same agreement in 1955. It broke the promises it made to the Western nations during previous meetings "at the summit" in Teheran and

[51]"Treason against the United States shall consist only in levying War against them, or in adhering to their enemies, giving them Aid and Comfort. . ."

[52]*Congressional Record*, vol. 74, p. 7049.

[53]David J. Dallin, quoted in *The Real Soviet Russia* (New Haven: Yale University Press, 1971), p. 71.

Yalta. It broke lend-lease agreements offered to it by the United States in order to keep Stalin from surrendering to the Nazis. It violated the charter of the United Nations. It keeps no international promises at all unless doing so is clearly advantageous to the Soviet Union.

[We] seriously doubt whether during the whole history of civilization any great nation has ever made as perfidious a record as this in so short a time.[54]

More recently in the 1970s and 1980s the Soviets have broken the SALT treaties and used the era of detente to develop an awe-inspiring weapons arsenal.

Consequently, the history of Soviet foreign relations from 1917 to the present suggests, for those who can interpret history, two conclusions:

1. The Soviets will not keep their word in any foreign agreement.
2. Their intent is self-admittedly aggressive, with world conquest as the ultimate goal.

The 1970s era of detente was a sham. Increased U.S.-Soviet trade, allegedly designed to lower tensions, was entirely contrary to historical observation and rational deduction. Mikhail Suslov, longtime Russian Communist Party theoretician, stated in 1972 that the U.S.-Soviet detente was temporary and that, so far as the Soviet Union is concerned, merely an interlude to gain strength for the next stage of the battle against "imperialist aggression." Suslov in 1972 repeated and reinforced Zinoviev's 1919 statement; there is no change of heart or direction.

The Soviet Record of Aggression

A review of the human cost of Soviet double-dealing emphasizes not only the risk we run by attempts to mellow Soviet statism, but the extreme seriousness of the actions of the deaf mute blindmen.

In every year since the Bolshevik Revolution the Soviets have murdered their own citizens for political reasons: that is, for alleged or real opposition to the Soviet state. The AFL-CIO has mapped Soviet forced labor camps. Moreover, in every year since 1917 the Soviets have attacked other countries or interfered massively in their internal affairs.[55]

The human cost of the Bolshevik Revolution and the ensuing civil war in Russia has been estimated at 7 million Russians. Between 1930 and 1950 more than 20 million Russians died in forced labor camps.

[54]U.S. Senate, Committee on the Judiciary, *Soviet Political Agreements and Results*, 4th printing (Washington, 1964).

[55]For the early years there is a State Department Staff report, *Interference of Representatives or Employees of the Soviet Government Abroad in the Internal Affairs of the Countries in Which They Are Stationed.*

Khrushchev personally supervised the massacre of more than 10,000 Ukrainians at Vinnitsa.

Soviet agents were in Spain before the Spanish Civil War of 1936 and unquestionably had some role in starting it (cost: 275,000 killed).

The supply of Soviet armaments to the Spanish Republic is known from material in the records of the German military attache at Ankara, Turkey.[56] Soviet arms shipments began in September 1936. Soviet intelligence agents, operating in Spain before the war broke out, were under General Ulansky, who was also responsible for logistics. In addition to supplies, the Soviets sent 920 military "advisers": 70 air force officers, 100 other officers (as early as September 1936), and 750 enlisted men. From September 1936 to March 1938 about 110 shiploads of Russian military supplies left Odessa en route to Spain, almost all from plants built by the deaf mute blindmen. Foreshadowing the situation when the USSR supplied Cuba and North Vietnam, only thirty-two of these ships were under the Soviet flag — and most of these Soviet-flag vessels were foreign-built. These 110 vessels carried the following armaments to Spain from the new Western-built Soviet plants:

Tanks and armored cars	731
Planes (mostly fighter aircraft)	242
Guns	707
Antiaircraft Guns	27
Trucks	1,386

What was the U.S. technical component of these arms?

The tanks sent to Spain in 1936 were based on British Vickers or U.S. Christie designs. Soviet aviation technology was mainly American (except for French Potez and Italian seaplane designs). The guns were Krupp, but the trucks were Ford, Hercules, and Brandt — all from plants built by American firms just five years previously.

After this, in 1937, Stalin's Red Army purge killed 30,000 — the cream of the Soviet military.

Two years later, in 1939, Russia attacked Finland. Cost: 273,000 Finns and Russians killed. In 1939 or 1940, the Soviets murdered 30,000 Polish officers at Katyn.

Persecution of Russians and the peoples of Eastern Europe continued after World War II, assisted by the British-American Operation Keelhaul.

In 1946 the Ukrainians tried unsuccessfully to fight for independence, after having fought the Germans for four years. One after another the East European peoples attempted to overthrow domestic communism,

[56]D.C. Warr, "Soviet Military Aid to the Spanish Republic in the Civil War 1936-1938," Slavonic and East European Review, June 1960, pp. 536-41. Also see: Uri Ra'anan, The USSR Arms the Third World (Cambridge, Mass.: MIT Press, 1956).

which survived only with Russian help and American inaction. In 1956 there was another Polish revolt and a major outbreak in Hungary in which 25,000 Hungarians and 7,000 Russians lost their lives.

In the early 1960s the Soviets began to look beyond their satellites, secure in the knowledge that the United States would not intervene to protect human rights in these countries. Then came the Cuban Missile Crisis of 1962: the missile-carrying Soviet ships had engines manufactured by Denmark. Then the airborne Congo adventure. Then the Vietnamese War, in which Soviet advisers, as in the Spanish Civil War in 1936, entered at an early stage. In 1965, the year 1,369 Americans were killed in South Vietnam, 2,500 Russian engineers and experts were at work in North Vietnam, and Russian arms were even then in widespread use by the Viet Cong and the North Vietnamese. This was also the year in which President Johnson decided to expand trade with the Soviets in the guise of "building bridges for peace."

The continuing crisis in the Middle East has been directly dependent upon the supply of Soviet arms to militant Arab countries and guerillas.

In the 1980s Soviet weapons and supplies keep wars going in Africa and Central America. Famine plagued Ethiopia is stocked with Soviet weapons. In Afghanistan we have a Soviet invasion of an independent country, Marxist Angola and Marxist Mozambique are supplied with Soviet weapons and advisers. In Angola the Soviets are in an unholy alliance with Chevron-Gulf. This oil multinational supplies the foreign exchange to pay the Cuban troops. The Gulf Cabinda installations are protected by Soviet and Cuban troops.

In Korea we have direct killing of Americans with Soviet weapons. The American casualty roll in the Korean War was 33,730 killed and 103,284 wounded. Of the 10,218 American prisoners taken by the Communist forces, only 3,746 returned to the United States: 21 men refused repatriation and 6,451 American servicemen are listed as "murdered or died."[57]

The 130,000-man North Korean Army, which crossed the South Korean border in June 1950, was trained, supported, and equipped by the Soviet Union, and included a brigade of Soviet T-34 medium tanks (with U.S. Christie suspensions).[58] The artillery tractors were direct metric copies of Caterpillar tractors. The trucks came from the Henry Ford-Gorki plant or the ZIL plant. The North Korean Air Force has 180 Yak planes built in plants with U.S. Lend-Lease equipment. These Yaks were later replaced by MiG-15s powered by Russian copies of Rolls-Royce jet engines sold to the Soviet Union in 1947.

[57] R. Ernest Dupuy and Trevor N. Dupuy, *Encyclopaedia of Military History* (New York: Harper & Row, 1970), p. 1219.

[58] Antony C. Sutton, *Western Technology and Soviet Economic Development, 1930-1945* (see Bibliography).

Between 1961 and 1964 the American casualty roll in Vietnam was relatively light, only 267 killed, and U.S.-Soviet trade was at a low level.

In 1965 the Soviets stepped up the flow of military supplies and equipment to North Vietnam. President Johnson stepped up the flow of technology to the Soviets. The American toll mounted rapidly, demonstrating the absurdity of the trade levels in peace agreement:

1965	1,369 killed	3,308 wounded
1966	5,008 killed	16,526 wounded
1967	9,378 killed	32,371 wounded
1968	14,592 killed	46,799 wounded

After President Nixon took office in 1969 and initiated detente with transfers of **military** technology, the American toll increased:

1969	9,414 killed	32,940 wounded
1970	4,422 killed	15,211 wounded
1971	1,380 killed	4,767 wounded
1972	300 killed	587 wounded

About 80 percent of the armaments and supplies for the Vietnamese War came from the Soviet Union. Yet a key part of President Nixon's policy was the transfer of technology to the USSR which aids Soviet war potential.

Soviet military aid has been fundamental for the North Viets. In September 1967 the Institute for Strategic Studies in London reported that the Soviets sent numbers of MiG-17 and MiG-21 fighters, Ilyushin-28 light bombers, transport aircraft, helicopters, 6,000 antiaircraft guns (one-half radar controlled), surface-to-air (guideline) missiles, 200-250 missile launchers, several thousand air defense machine guns, and a training mission of about 1,000 men to North Vietnam.

This aid was confirmed in April 1967 by former Assistant Secretary of Defense John T. McNaughton, i.e. the Soviets supplied the "sophisticated equipment in the field of antiaircraft defense." Loss of 915 U.S. planes over North Vietnam between February 1965 (the date of the first U.S. air operations over North Vietnam) and the bombing halt of November 1, 1968 testifies to the accuracy and utility of the Soviet equipment. After President Nixon took office in January 1969 and expanded technical transfers, losses mounted, a total of more than 4,000 U.S. aircraft by the end of 1972.

Support by the Soviet Union for North Vietnamese aggression in South Vietnam was no secret. Brezhnev, on his visit to Bulgaria on May 12, 1967, demonstrated the solidarity of supposedly polycentralist-socialist East Europe on the question of Vietnam:

> You know well, comrades, that the Soviet Union is rendering great economic, military and political assistance to fighting Viet-

nam. This assistance is merged with the assistance coming from Bulgaria and other fraternal socialist countries. We are rendering it in response to a command of the heart, as people reared by the Communist party in a spirit of proletarian internationalism, in a spirit of high understanding of class tasks. And let the aggressors know this: fighting Vietnam will never be left without the help of its true friends. Our answer has been and will continue to be commensurate with the requirements of an effective rebuff to the unbridled imperialist interventionists.[59]

The Soviet Union provided both the military and the economic means for the North Vietnamese invasion of South Vietnam. At the key juncture of the Vietnamese War, the Soviets were not only rapidly increasing their supplies, but were receiving in return less than one-seventh the value of the supplies in Vietnamese products. The balance was Soviet "Lend-Lease" for the takeover of South Vietnam.

The Soviet Union has truly been the "arsenal for revolution" in Vietnam, and as Shirley Sheibla wrote in *Barron's Weekly*, the United States has been the "arsenal for communism" in the Soviet Union."[60]

As the material presented in this book shows, the "arsenal for revolution" was built by Western firms and has been kept in operation with "peaceful trade." When all the rhetoric about "peaceful trade" is boiled out, it comes down to a single inescapable fact: the guns, the ammunition, the weapons, the transportation systems that killed Americans in Korea and Vietnam came from the American-subsidized economy of the Soviet Union. The trucks that carried these weapons down the Ho Chi Minh trail came from American-built plants. The ships that carried the supplies to Sihanoukville and Haiphong and later to Angola and Nicaragua came from NATO allies and used propulsion systems that our State Department could have kept out of Soviet hands — indeed, the Export Control Act and the Battle Act, ignored by State, required exactly such action. The technical capability to wage the Korean and Vietnamese wars originated **on both sides** in Western, mainly American, technology, and the political illusion of "peaceful trade" promoted by the deaf mute blindmen was the carrier for this war-making technology.

As U.S. casualties in Vietnam mounted, the lessons of history were clear for those with eyes to see — reduce trade with the USSR and all suppliers to North Vietnam, and so provide an incentive for the other side to decelerate the conflict. (This is not hindsight; the writer made this argument, in print, in the mid-1960s and throughout the 1970s. See Appendix C). Both the Johnson and Nixon administrations irrationally

[59]*Pravda*, May 13, 1967.
[60]Jan. 4., 1971.

and illogically chose to expand trade — the carrier for the technology required to fuel the North Vietnamese side of the war — and so voted to continue the war.

What is less obvious is that all Administrations have been under heavy political pressure from the deaf mute blindmen to adopt this suicidal policy.

Are the Deaf Mute Blindmen Guilty of Treason?

The Soviet record is clear. Since 1917 the Soviets have in philosophy and action held that the United States is the main enemy. The Soviets talk as if the United States is an enemy. And Soviets consistently act as if the United States is the main enemy. We also require a $300 billion a year defense budget to protect ourselves from the Soviets.

What does the U.S. Constitution say about treason? Here is Article III, Section 3:

United States Constitution
ARTICLE III, SECTION 3

TREASON AGAINST THE UNITED STATES SHALL CONSIST ONLY IN LEVYING WAR AGAINST THEM, OR IN ADHERING TO THEIR ENEMIES, GIVING THEM AID AND COMFORT. NO PERSON SHALL BE CONVICTED OF TREASON UNLESS ON THE TESTIMONY OF TWO WITNESSES TO THE SAME OVERT ACT, OR IN CONFESSION IN OPEN COURT.

THE CONGRESS SHALL HAVE POWER TO DECLARE THE PUNISHMENT OF TREASON, BUT NO ATTAINDER OF TREASON SHALL WORK CORRUPTION OF BLOOD, OR FORFEITURE, EXCEPT DURING THE LIFE OF THE PERSON ATTAINTED.

Treason is defined in the Constitution as giving any enemy of the United States "aid and comfort." Does the record described in the previous chapters constitute "aid and comfort"?

Obviously the record reflects considerably more than "aid and comfort." The Soviets would have no effective modern military establishment without the assistance rendered by the deaf mute blindmen — on credit at that.

The difference between 1985 and 1974 when we declined to suggest treason as an explanation is that there has now been a full decade for the deaf mute blindmen to acquaint themselves with the facts, and draw back from those actions that could be interpreted as treason.

On the contrary, not only have the subsidies expanded, but they have openly included military technology. Henry Kissinger, Chase Manhattan Bank and the suppliers of the Kama River plant KNEW the plant had potential to manufacture **military** trucks. Henry Kissinger and the Bryant Chucking Grinder Company KNEW the Centalign-B machine was used to machine the races for precision ball bearings used in Soviet missiles. It is this fact of CLEAR KNOWLEDGE that pushes us towards a conclusion of treason. If a person willfully, i.e. **knowingly and deliberately supplies military technology to any enemy,** then he must under any definition of treason be guilty of that crime.

To conclude: if "aid and comfort" to any enemy is treason, then the deaf mute blindmen are guilty of treason.

We now have the formidable task of bringing these gentlemen to the bar of justice to publicly answer for their private and concealed actions.

Appendix A:
Exchange of Letters with Department of Defense, 1971

These letters contradict the statement made by William J. Casey, Director of Central Intelligence Agency, that in 1985 the Administration had only "recently" learned of the impact of our technology on the Soviet armaments industry. Washington was alerted 15 years ago.

September 15, 1971

Dr. N.F. Wikner
Special Assistant for Threat Assessment
Director of Defense Research and Engineering
Department of Defense
The Pentagon
Washington D.C. 20301

Dear Dr. Wikner:

I am in receipt of a letter from the office of the Director of Defense Research and Engineering (signed Eberhardt Rechtin) in regard to an inquiry made by me to the Secretary concerning technical information for a projected book WESTERN TECHNOLOGY AND THE SOVIET ARMAMENTS INDUSTRY.

In response to this letter, which suggests that I should contact you, I enclose:

    a. a list of the information desired

    b. a very approximate and preliminary outline of the structure of the book.

I shall, of course be more than happy to clarify any points that may arise in consideration of my request.

With best wishes,

Very sincerely,
Antony C. Sutton
Research Fellow

Details of Information Requested
The information required is as follows:

a) Detailed technical and engineering data on Soviet weapons systems from 1945 to date, in the form of technical handbooks or reports (maintenance or servicing handbooks are adequate but less valuable). In Russian or English with diagrammatic layouts, cutaways, technical specifications of materials used and metallurgical analyses.

These are needed for the production models in each weapons series from 1945 to date. For example, MEDIUM TANKS: data is needed on the T 34, T 54 and T 62; but data is *not* needed on development models, such as T 44 or variants of main production models such as T 34/76B (a turret variant of the main production model T 34).

The weapons spectrum for which this range of data is required are the production models of:

tanks (heavy, medium and light), armored cars, self propelled guns, trucks and tractors, guns of all types (from tank guns down to hand guns), ammunition, planes, naval craft, rocket launchers, missiles. In other words the standard models in the broad weapons spectrum.

I am not interested in the more esoteric weapons under development (such as lasers), or weapons developed and abandoned, but only those systems which constitute (or have historically constituted) the *main* threat to the Free World.

b) A detailed listing of the inputs required to manufacture each of the above Soviet weapons including if possible the chemical or physical specifications of material inputs, quantities of inputs per weapon, and model numbers and types of equipment.

Categories (a) and (b) are required in order to determine *how* Soviet weapons are manufactured and *what* material inputs are used.

c) Reports or raw data on the use by the Soviets of Western technology in weapons systems and general military production Equipment rosters of Soviet armament plants, i.e. their machinery inventories, (these will identify use of Western machines).

d) Reports or raw data on Soviet manufacture of propellants, explosives, military clothing and instrumentation and computers; the process used, outputs, names of plants.

For example: I would like to know the types of explosives produced by the Soviets and either the chemical anlysis or sufficient information to determine an approximate analysis. I am not interested in the military aspects i.e. the explosive force or characteristics of the explosion, only *what* is being produced and *how* it is being produced.

e) Material on the conversion of a civilian industrial base to a military base; the U.S. experience in World War II and Korea; the problems of conversion, time required, adaptability of a civilian plant to military output.

f) Details of the important Export Control cases (both under the Export Control Act and CoCom in Paris) where DOD has argued against export of technology or equipment items to the Soviet Union or to other countries where there was a possibility of transfer to the USSR. These would include for example, the Transfermatic Case of 1961 and the Ball Bearings case of about the same date. From the mid 1950's down to the present time.

g) Equipment lists (by model number, not necessarily quantities) of North Viet and Viet Cong forces.

In general I am *not* interested in the quantitative aspects (i.e. how many they have of a particular weapon) nor in military characteristics (i.e. ballistic properties, operating characteristics etc.).

On the other hand I am interested in qualitative aspects, particularly knowing *how* weapons are produced and the material and equipment inputs used to produce these weapons. Whether the weapons and materials produced are militarily or economically efficient is of little concern for this study.

February 18, 1972

Dr. N.F. Wikner
Special Assistant for Threat Assessment
Director of Defense Research and Engineering
Washington D.C. 20301

Dear Dr. Wikner:

This refers to my letter of September 15, 1971 concerning my request for technical information on the transfer of technology to the Soviet Union.

My understanding is that the Department has a suitable data base and that according to Mr. Eberhardt Rechtin's letter of September 9: "there is a systematic and continuous effort to declassify all pertinent information."

Accordingly I submitted with my letter a detailed statement of the information desired.

In the absence of any reply or acknowledgement from the Department in the elapsed five months, would it now be accurate for me to assume that there is no desire to pursue the question further?

Sincerely,
Antony C. Sutton

Appendix B:

Testimony of the Author Before Subcommittee VII of the Platform Committee of the Republican Party at Miami Beach, Florida, August 15, 1972, at 2:30 P. M.

This appendix contains the testimony presented by the author before the Republican Party National Security Subcommittee at the 1972 Miami Beach convention. The author's appearance was made under the auspices of the American Conservative Union; the chairman of the subcommittee was Senator John Tower of Texas.

Edith Kermit Roosevelt subsequently used this testimony for her syndicated column in such newspapers as the *Union Leader* (Manchester, N.H.). Both major wire services received copies from the American Conservative Union; they were not distributed. Congressman John G. Schmitz then arranged for duplicate copies to be hand-delivered to both UPI and AP. The wire services would not carry the testimony although the author is an internationally known academic researcher with three books published at Stanford University, and a forthcoming book from the U.S. Naval Institute.

The testimony was later reprinted in full in *Human Events* (under the title of "The Soviet Military-Industrial Complex") and *Review of the News* (under the title of "Suppressed Testimony of Antony C. Sutton"). It was also reprinted and extensively distributed throughout the United States by both the American party and the Libertarian party during the 1972 election campaign.

The following is the text of this testimony as it was originally presented in Miami Beach and made available to UPI and AP:

The Soviet Military–Industrial Complex

The information that I am going to present to you this afternoon *is* known to the Administration.

The information is probably *not* known to the Senator from South Dakota or his advisers. And in this instance ignorance may be a blessing in disguise.

I am not a politician. I am not going to tell you what you want to hear. My job is to give you facts. Whether you like or dislike what I say doesn't concern me.

I am here because I believe—and Congressman Ashbrook believes—that the American public should have these facts.

I have spent ten years in research on Soviet technology. What it is—what it can do—and particularly where it came from. I have published three books and several articles summarizing the work.

It was privately financed. But the results have been available to the Government. On the other hand I have had major difficulties with U.S. Government censorship.

I have 15 minutes to tell you about this work.

In a few words: there is no such thing as Soviet technology.

Almost all—perhaps 90–95 percent—came directly or indirectly from the United States and its allies. In effect the United States and the NATO countries have built the Soviet

Union. Its industrial *and* its military capabilities. This massive construction job has taken 50 years. Since the Revolution in 1917. It has been carried out through trade and the sale of plants, equipment and technical assistance.

Listening to Administration spokesmen—or some newspaper pundits—you get the impression that trade with the Soviet Union is some new miracle cure for the world's problems.

That's not quite accurate.

The idea that trade with the Soviets might bring peace goes back to 1917. The earliest proposal is dated December 1917— just a few weeks after the start of the Bolshevik Revolution. It was implemented in 1920 while the Bolsheviks were still trying to consolidate their hold on Russia. The result was to guarantee that the Bolsheviks held power: they needed foreign supplies to survive.

The history of our construction of the Soviet Union has been blacked out—much of the key information is still classified— along with the other mistakes of the Washington bureaucracy.

Why has the history been blacked out?

Because 50 years of dealings with the Soviets has been an economic success for the USSR and a political failure for the United States. It has not stopped war, it has not given us peace.

The United States is spending $80 billion a year on defense against an enemy built by the United States and West Europe.

Even stranger, the U.S. apparently wants to make sure this enemy remains in the business of being an enemy.

Now at this point I've probably lost some of you. What I . have said is contrary to everything you've heard from the intellectual elite, the Administration, and the business world, and numerous well-regarded Senators—just about everyone.

Let me bring you back to earth.

First an authentic statement. It's authentic because it was part of a conversation between Stalin and W. Averell Harriman. Ambassador Harriman has been prominent in Soviet trade since the 1930's and is an outspoken supporter of yet more trade. This is what Ambassador Harriman reported back to the State Department at the end of World War II:

"Stalin paid tribute to the assistance rendered by the United States to Soviet industry before and during the War. Stalin* said that about two-thirds of all the large industrial enterprises in the Soviet Union has been built with the United States' help or technical assistance."

I repeat: "two-thirds of all the large industrial enterprises in the Soviet Union had been built with the United States' help or technical assistance."

Two-thirds.

Two out of three.

Stalin could have said that the other one-third of large industrial enterprises were built by firms from Germany, France, Britain and Italy.

Stalin could have said also that the tank plants, the aircraft plants, the explosive and ammunition plants originated in the U.S.

That was June 1944. The massive technical assistance continues right down to the present day.

Now the ability of the Soviet Union to create any kind of military machine, to ship missiles to Cuba, to supply arms to North Vietnam, to supply arms for use against Israel—all this depends on its domestic industry.

In the Soviet Union about three-quarters of the military budget goes on purchases from Soviet factories.

This expenditure in Soviet industry makes sense. No Army has a machine that churns out tanks. Tanks are made from alloy steel, plastics, rubber and so forth. The alloy steel, plastics and rubber are made in Soviet factories to military specifications. Just like in the United States.

Missiles are not produced on missile-making machines. Missiles are fabricated from aluminum alloys, stainless steel, electrical wiring, pumps and so forth. The aluminum, steel, copper wire and pumps are also made in Soviet factories.

In other words the Soviet military gets its parts and materials

* He, in original.

from Soviet industry. There is a Soviet military-industrial complex just as there is an American military-industrial complex.

This kind of reasoning makes sense to the man in the street. The farmer in Kansas knows what I mean. The salesman in California knows what I mean. The taxi driver in New York knows what I mean. But the policy makers in Washington do not accept this kind of common sense reasoning, and never have done.

So let's take a look at the Soviet industry that provides the parts and the materials for Soviet armaments: the guns, tanks, aircraft.

The Soviets have the largest iron and steel plant in the world. It was built by McKee Corporation. It is a copy of the U.S. Steel plant in Gary, Indiana.

All Soviet iron and steel technology comes from the U.S. and its allies. The Soviets use open hearth, American electric furnaces, American wide strip mills, Sendzimir mills and so on—all developed in the West and shipped in as peaceful trade.

The Soviets have the largest tube and pipe mill in Europe—one million tons a year. The equipment is Fretz-Moon, Salem, Aetna Standard, Mannesman, etc. Those are not Russian names.

All Soviet tube and pipe making technology comes from the U.S. and its allies. If you know anyone in the space business ask them how many miles of tubes and pipes go into a missile.

The Soviets have the largest merchant marine in the world—about 6,000 ships. I have the specifications for each ship.

About two-thirds were built outside the Soviet Union.

About four-fifths of the engines for these ships were also built outside the Soviet Union.

There are no ship engines of Soviet design. Those built *inside* the USSR are built with foreign technical assistance. The Bryansk plant makes the largest marine diesels. In 1959, the Bryansk plant made a technical assistance agreement with Burmeister & Wain of Copenhagen, Denmark, (a NATO ally), approved as peaceful trade by the State Dept. The ships that carried Soviet missiles to Cuba ten years ago used these same Burmeister and Wain engines. The ships were in the POLTAVA class. Some have Danish engines made in Denmark and some have Danish engines made at Bryansk in the Soviet Union.

About 100 Soviet ships are used on the Haiphong run to carry Soviet weapons and supplies for Hanoi's annual aggression. I was able to identify 84 of these ships. None of the main engines in these ships was designed and manufactured inside the USSR.

All the larger and faster vessels on the Haiphong run were built outside the USSR.

All shipbuilding technology in the USSR comes directly or indirectly from the U.S. or its NATO allies.

Let's take one industry in more detail: motor vehicles.

All Soviet automobile, truck and engine technology comes from the West: chiefly the United States. In my books I have listed each Soviet plant, its equipment and who supplied the equipment. The Soviet military has over 300,000 trucks—all from these U.S. built plants.

Up to 1968 the largest motor vehicle plant in the USSR was at Gorki. Gorki produces many of the trucks American pilots see on the Ho Chi Minh trail. Gorki produces the chassis for the GAZ-69 rocket launcher used against Israel. Gorki produces the Soviet jeep and half a dozen other military vehicles.

And Gorki was built by the Ford Motor Company and the Austin Company—as peaceful trade.

In 1968 while Gorki was building vehicles to be used in Vietnam and Israel further equipment for Gorki was ordered and shipped from the U.S.

Also in 1968 we had the so-called "FIAT deal"—to build a plant at Volgograd three times bigger than Gorki. Dean Rusk and Walt Rostow told Congress and the American public this was peaceful trade—the FIAT plant could not produce military vehicles.

Don't let's kid ourselves. *Any* automobile manufacturing plant can produce military vehicles. I can show anyone who is interested the technical specification of a proven military vehicle (with cross-country capability) using the same capacity engine as the Russian FIAT plant produces.

The term "FIAT deal" is misleading. FIAT in Italy doesn't make automobile manufacturing equipment—FIAT plants in Italy have U.S. equipment. FIAT *did* send 1,000 men to Russia for erection of the plant—but over half, perhaps well over half, of the equipment came from the United States. From Gleason, TRW of Cleveland and New Britain Machine Co.

So in the middle of a war that has killed 46,000 Americans (so far) and countless Vietnamese with Soviet weapons and supplies, the Johnson Administration doubled Soviet auto output.

And supplied false information to Congress and the American public.

Finally, we get to 1972 under President Nixon.

The Soviets are receiving now—today, equipment and technology for the largest heavy truck plant in the world: known as the Kama plant. It will produce 100,000 heavy ten-ton trucks per year—that's more than ALL U.S. manufacturers put together.

This will also be the largest plant in the world, *period*. It will occupy 36 square miles.

Will the Kama truck plant have military potential?

The Soviets themselves have answered this one. The Kama truck will be 50 per cent more productive than the ZIL-130 truck. Well, that's nice, because the ZIL series trucks are standard Soviet army trucks used in Vietnam and the Middle East.

Who built the ZIL plant? It was built by the Arthur J. Brandt Company of Detroit, Michigan.

Who's building the Kama truck plant? That's classified "secret" by the Washington policy makers. I don't have to tell you why.

The Soviet T-54 tank is in Vietnam. It was in operation at Kontum, An Loc, and Hue a few weeks ago. It is in use today in Vietnam. It has been used against Israel.

According to the tank handbooks the T-54 has a Christie type suspension. Christie was an American inventor.

Where did the Soviets get a Christie suspension? Did they steal it?

No, sir! They bought it. They bought it from the U.S. Wheel Track Layer Corporation.

However this Administration is apparently slightly more honest than the previous Administration.

Last December I asked Assistant Secretary Kenneth Davis of the Commerce Department (who is a mechanical engineer by training) whether the Kama trucks would have military capability. In fact I quoted one of the Government's own inter-agency reports. Mr. Davis didn't bother to answer but I did get a letter from the Department and it was right to the point. Yes! we know the Kama truck plant has military capability, we take this into account when we issue export licenses.

I passed these letters on to the press and Congress. They were published.

Unfortunately for my research project, I also had pending with the Department of Defense an application for declassification of certain files about our military assistance to the Soviets.

This application was then abruptly denied by DOD.

It will supply military technology to the Soviets but gets a little uptight about the public finding out.

I can understand that.

Of course, it takes a great deal of self confidence to admit you are sending factories to produce weapons and supplies to a country providing weapons and supplies to kill Americans, Israelis and Vietnamese. In writing. In an election year, yet.

More to the point—by what authority does this Administration undertake such policies?

Many people—as individuals—have protested our suicidal policies. What happens? Well, if you are in Congress—you probably get the strong arm put on you. The Congressman who inserted my research findings into the Congressional Record suddenly found himself with primary opposition. He won't be in Congress next year.

If you are in the academic world—you soon find it's OK to protest U.S. assistance to the South Vietnamese but never, never protest U.S. assistance to the Soviets. Forget about the Russian academics being persecuted—we mustn't say unkind things about the Soviets.

If you press for an explanation what do they tell you?

First, you get the Fulbright line. This is peaceful trade. The Soviets are powerful. They have their own technology. It's a way to build friendship. It's a way to a new world order.

This is demonstrably false.

The Soviet tanks in An Loc are not refugees from the Pasadena Rose Bowl Parade.

The "Soviet" ships that carry arms to Haiphong are not peaceful. They have weapons on board, not flower children or Russian tourists.

Second, if you don't buy that line you are told, "The Soviets are mellowing." This is equally false.

The killing in Israel and Vietnam with Soviet weapons doesn't suggest mellowing, it suggests premeditated genocide. Today—*now*—the Soviets are readying more arms to go to Syria. For what purpose? To put in a museum?

No one has ever presented evidence, hard evidence that trade leads to peace. Why not? Because there *is* no such evidence. It's an illusion.

It is true that peace leads to trade. But that's not the same thing. You first need peace, then you trade. That does not mean if you trade you will get peace.

But that's too logical for the Washington policy makers and it's not what the politicians and their backers want anyway.

Trade with Germany doubled before World War II. Did it stop World War II?

Trade with Japan increased before World War II. Did it stop World War II?

What was in this German and Japanese trade? The same means for war that we are now supplying the Soviets. The Japanese Air Force after 1934 depended on U.S. technology. And much of the pushing for Soviet trade today comes from the same groups that were pushing for trade with Hitler and Tojo 35 years ago.

The Russian Communist Party is not mellowing. Concentration camps are still there. The mental hospitals take the overload. Persecution of the Baptists continues. Harassment of Jews continues, as it did under the Tsars.

The only mellowing is when a Harriman and a Rockefeller get together with the bosses in the Kremlin. That's good for business but it's not much help if you are a G.I. at the other end of a Soviet rocket in Vietnam.

I've learned something about our military assistance to the Soviets.

It's just not enough to have the facts—these are ignored by the policy makers.

It's just not enough to make a common sense case—the answers you get defy reason.

Only one institution has been clearsighted on this question. From the early 1920's to the present day only one institution has spoken out. That is the AFL-CIO.

From Samuel Gompers in 1920 down to George Meany today, the major unions have consistently protested the trade policies that built the Soviet Union.

Because union members in Russia lost their freedom and union members in the United States have died in Korea and Vietnam.

The unions know—and apparently care.

No one else cares. Not Washington. Not big business. Not the Republican Party.

And 100,000 Americans have been killed in Korea and Vietnam—by our own technology.

The only response from Washington and the Nixon Administration is the effort to hush up the scandal.

These are things not to be talked about. And the professional smokescreen about peaceful trade continues.

The plain fact—if you want it—is that irresponsible policies have built us an enemy and maintain that enemy in the business of totalitarian rule and world conquest.

And the tragedy is that intelligent people have bought the political double talk about world peace, a new world order and mellowing Soviets.

I suggest that the man in the street, the average taxpayer-voter thinks more or less as I do. You do not subsidize an enemy.

And when this story gets out and about in the United States, it's going to translate into a shift of votes. I haven't met one man in the street so far (from New York to California) who goes along with a policy of subsidizing the killing of his fellow Americans. People are usually stunned and disgusted.

It requires a peculiar kind of intellectual myopia to ship supplies and technology to the Soviets when they are instrumental in killing fellow citizens.

What about the argument that trade will lead to peace? Well, we've had U.S.-Soviet trade for 52 years. The 1st and 2nd Five Year Plans were built by American companies. To continue a policy that is a total failure is to gamble with the lives of several million Americans and countless allies.

You can't stoke up the Soviet military machine at one end and then complain that the other end came back and bit you. Unfortunately, the human price for our immoral policies is not paid by the policy maker in Washington. The human price is paid by the farmers, the students and working and middle classes of America.

The citizen who pays the piper is not calling the tune—he doesn't even know the name of the tune.

Let me summarize my conclusions:

One: trade with the USSR was started over 50 years ago under President Woodrow Wilson with the declared intention of mellowing the Bolsheviks. The policy has been a total and costly failure. It has proven to be impractical—this is what I would expect from an immoral policy.

Two: we have built ourselves an enemy. We keep that self-declared enemy in business. This information has been blacked out by successive Administrations. Misleading and untruthful statements have been made by the Executive Branch to Congress and the American people.

Three: our policy of subsidizing self-declared enemies is neither rational nor moral. I have drawn attention to the intellectual myopia of the group that influences and draws up foreign policy. I suggest these policies have no authority.

Four: the annual attacks in Vietnam and the war in the Middle East were made possible only by Russian armaments and our past assistance to the Soviets.

Five: this worldwide Soviet activity is consistent with Communist theory. Mikhail Suslov, the party theoretician, recently stated that the current detente with the United States is temporary. The purpose of the detente, according to Suslov, is to give the Soviets sufficient strength for a renewed assault on the West. In other words, when you've finished building the Kama plant and the trucks come rolling off—watch out for another Vietnam.

Six: internal Soviet repression continues—against Baptists, against Jews, against national groups and against dissident academics.

Seven: Soviet technical dependence is a powerful instrument for world peace if we want to use it.

So far it's been used as an aid-to-dependent-Soviets welfare program. With about as much success as the domestic welfare program.

Why should they stop supplying Hanoi? The more they stoke up the war the more they get from the United States.

One final thought.

Why has the war in Vietnam continued for four long years under this Administration?

With 15,000 killed under the Nixon Administration?

We can stop the Soviets and their friends in Hanoi anytime we want to.

Without using a single gun or anything more dangerous than a piece of paper or a telephone call.

We have Soviet technical dependence as an instrument of world peace. The most humane weapon that can be conceived.

We have always had that option. We have never used it.

Appendix C:

Letter from William C. Norris, Chairman of Control Data Corporation to Congressman Richard T. Hanna, 1973.

Control Data Corp.
Minneapolis, Minn., December 19, 1973.

Hon. Richard T. Hanna,
Rayburn House Office Building,
Washington, D.C.

My Dear Congressman Hanna: On Wednesday, December 5, 1973, testimony was given before the Subcommittee on International Cooperation in Science and Space of the House Science & Astronautics Committee by Mr. Benjamin Schemmer, Editor, *Armed Forces Journal International.* This testimony included the statement that Control Data Corporation had advanced the status of Soviet Computer technology by fifteen years with the sale of a Control Data 6200 computer.

Such a statement regarding transfer of technology to the USSR is simply not factual and we are prepared to correct that misstatement as well as other incorrect and misleading references to Control Data's activities with the USSR at the pleasure of your Committee. Meanwhile we respectfully request the consideration of the following.

We have offered to the Socialist countries only standard commercial computers, and these offerings have been in full compliance with the export control and administrative directives of the Department of Commerce.

The statement regarding a proposed sale of the CYBER computer is thoroughly confused. CYBER is a generic name denoting a line of computers. The least powerful model is the Control Data 6200 which is installed at the Dubna Nuclear Research facility near Moscow. Another is the CYBER 76 which is the most powerful and appears to be the model Mr. Schemmer is referencing. At the appropriate time we will seek advisory opinions and submit to the government export license requests for approval for applications in such areas as weather forecasting, simulation in the Worldwide Weather Watch Program, in econometric modeling and in education. Competition from West Europe and Japan will be expected to address these applications with the same kind of technologies that we offer.

Many persons, including some of the witnesses before your Committee, mistake the offering for sale of old or even current state of the art hardware for transfer of advanced technology. This is not unusual because in many cases it is difficult for those who are not technically well-informed to distingush advanced computer technology. Attempting to treat such a broad and complex subject in an informative and accurate manner in a brief statement results in inaccurate, unsubstantiated and misinterpreted inferences and conclusions. For example, it is possible that a computer with a slower arithmetic/logic unit and a very large memory may represent more "advanced" technology when measured by the size of the problem to be solved than a faster machine with less memory. Also, many minicomputers contain circuit technology more advanced than that of more powerful computers. For this reason it is not possible to properly respond to Mr. Schemmer's remarks in a short letter, however the following are examples of those which need to be carefully examined before conclusions are reached.

All countries including the Socialists have a substantial base of computer hardware technology on which to build further advances in the state of the art. The major strength of the U.S. in computer technology is its ability to market superior cost/performance computer systems for a wide range of applications. This does not mean that for any given application or group of applications, another country cannot build the equivalent as far as performance is concerned or even exceed what the United States has available. Also, there is no evidence to my knowledge that the USSR has ever been prevented from carrying out a military project because of the lack of adequate computer technology.

Further advances in hardware are less significant than are software advances for applying computers. We believe that the United States stands to gain significantly from transfer of Soviet knowledge in the basic sciences. The USSR has many more scientists and engineers than do we, and the better ones in Russia have concentrated on the theoretical fields, such as physics, chemistry and mathematics — the latter discipline being of particular value in the design of logic and software in the computer field.

We respectfully request that your Committee review the above points and consider incorporating them into the record. We would be pleased to have the privilege of appearing before your Committee to give you our more detailed views on these potential relationships with the Socialist countries and in stating our reasons in support of Administration and Congressional trade initatives and objectives.

Sincerely yours,

William C. Norris,
Chairman of the Board

Appendix D:

Letter from Fred Schlafly to friends and supporters of American Council for World Freedom, dated April 1978, asking to mail "Yellow Cards" of protest to William Norris.

Dear ACWF Supporter:

I need just a few seconds of your time right now.

I need you to sign and mail the 2 postcards I've enclosed for your personal use.

Since you're one of the best friends the American Council for World Freedom has, I'm sure I'm not asking too much of you.

Please let me explain these 2 postcards and tell you why it's **crucial** that you sign and mail them **today.**

The yellow postcard is addressed to Mr. William Norris, Chairman of the Control Data Corporation, one of the world's biggest and most advanced computer companies.

That yellow postcard calls on Mr. Norris and Control Data to stop selling computer technology to Communist Russia and its satellites . . .

. . . . Technology that our Communist enemies are using to gain military superiority over the U.S.

Once before, Mr. Norris tried to sell our best and most advanced computer to the Soviet Union. Only an all-out, last minute effort by over 300 patriotic Congressmen stopped the sale of this highly valuable computer to the Russians.

Now Mr. Norris and Control Data are trying the same sell-out to Russia **again.** You and I have got to put a stop to it.

And just to make sure that Mr. Norris **personally** sees **your** postcard, I've adressed the card to his home address in St. Paul, Minnesota.

The blue postcard is addressed to Mr. J. Fred Bucy, President of Texas Instruments.

That blue postcard **praises** Mr. Bucy and Texas Instruments for refusing to sell American technology to our Communist enemies.

You probably know from reading your anti-Communist magazines and newspapers that a lot of big, U.S. corporations are making huge profits selling their products to Communist Russia.

Now I'm not talking about selling products like soft drinks and clothing to the Communists.

I'm talking about U.S. companies . . . like Control Data . . . that sell our Communist enemies computers, shipbuilding equipment, and jet airplanes . . .

. . . Technology that Communist Russia is using to turn itself into the world's number one military superpower.

My friend, let me tell you just how critical a problem we really face. There are over 60 big U.S. companies selling U.S. technological secrets to the Soviet Union . . .

. . . Companies like Union Carbide, General Electric, Armco Steel, Bryant Chucking Grinder, which sells the Communists the ball bearings they need for their attack missiles . . . and Control Data.

So why is ACWF only going after Control Data, and not these other companies, for selling out U.S. military superiority to the Communists?

The answer is,

> Control Data is absolutely, far and away, the biggest offender when it comes to selling out U.S. technology and actively aiding the Soviet military power grab.

There's no doubt in my mind that if we ACWF Supporters force Control Data to stop building the Soviet war machine, then other sell-out companies like General Electric and Union Carbide will also stop aiding and abetting our Communist enemies.

That's why the American Council for World Freedom has set up its Task Force on Strategic Trade.

Look at what ACWF has already accomplished against Control Data:

- Just recently, we had a very large, enormously successful press conference on Capitol Hill, announcing our fight against Control Data's sell-out of U.S. computer secrets to the Communists. Over 75 reporters and Congressmen were there!

  (You might want to look at the enclosed note from General Daniel Graham which lists some of the Congressmen, newspapers, magazines, and wire services at our press conference.)

- ACWF just published a devastating expose of how Control Data and other sell-out companies are turning America into a second-rate power by helping build the Soviet war machine.

- We gave each reporter and Congressman at our press conference a copy of our study, "The Strategic Dimension of East-West Trade" by Dr. Miles Costick, one of America's top experts on foreign affairs and strategic trade.

ACWF is off to a fast start against Control Data, but there's no use kidding ourselves — we've got a big fight ahead of us.

Here's what needs to be done right away:

1.) We've got to flood Control Data and its Chairman, Mr. William Norris, with postcards and letters demanding "No More U.S. Help for the Soviet War Machine."

You can help do that by signing and mailing your postcard to Mr. Norris . . . today.

2.) We've got to support and encourage companies like Texas Instruments when they show the guts and have the backbone to say "NO" to Communist Russia.

Will you help do that by mailing your postcard to Mr. J. Fred Bucy, President of Texas Instruments . . . today?

3.) We've got to print and distribute 28,959 copies of Dr. Costick's brilliant expose, "The Strategic Dimension of East-West Trade."

We must get this expose into the hands of America's 10,286 newspaper editors, 9,414 magainze editors, 9,309 TV and radio news editors.

4.) And finally, we've got to write to tens of thousands of other Americans . . . 50 thousand in May alone . . . alerting them to Control Data and similar sell-outs of America's scientific secrets. getting them active in ACWF's fight to keep America the #1 military power on earth.

I know I don't have to tell you that I'm talking about an aggressive, expensive program.

It will cost $28,090 to print 28,959 copies of ACWF's Control Data expose and distribute tham to 28,959 newspaper, magazine, radio and TV editors.

And it will cost $13,253 to write to fifty thousand Americans in May.

That's a total of $41,343 ACWF urgently needs. We simply don't have it.

Will you help ACWF again, with as generous a gift as you can afford?

What I'm asking you for is your contribution . . . before May 15th . . . of at least $25, but hopefully more — perhaps even $50, $100, $250, $500 or $1,000 if you can afford it.

I know I ask an awful lot of you, but I don't have anyone else to turn to.

So please, when you mail your 2 postcards, won't you use the enclosed reply envelope to rush ACWF your support of $25, $50, $100, $250, $500 or $1,000?

Gratefully,
Fred Schlafly

P.S. Remember, when you mail that yellow postcard protesting Control Data's military sell-out of the U.S. to Russia, your card **will** be read by Control Data Boss William Norris because it's addressed to his home.

Letter from William C. Norris to each "Yellow Card Sender," dated May 5, 1978.

CONTROL DATA
CORPORATION

·May 5, 1978

Dear Yellow Card Sender:

You are grossly unfair, woefully ignorant or being led around by the nose. I suspect it is the latter, otherwise you wouldn't have signed a card with a canned message, along with a request to contribute money to the author — apparently a Mr. Schlafly of American, Council for World Freedom.

Why didn't you write and ask about Control Data's position on trade with the Communists before signing a form card? You obviously have little knowledge about Control Data's position or activities, nor does Mr. Schlafly. He has not contacted anyone in Control Data. I would have furnished him information or I would have been glad to respond to a letter from you asking for information. Then if you still disagreed strongly — so be it — at least you would have acted sensibly and fairly by first getting adequate information. Then you probably wouldn't have resorted to the use of such words as "disgrace" and "outrage", because that is akin to the Communist procedure of not fairly weighing the evidence.

You obviously are deeply concerned about the Soviet military threat, and I certainly understand that, because I share that concern. I think about the subject a great deal and several points are clear, including that:

1. The U.S. should maintain a very strong military capability. Control Data's products and services contribute significantly to our nation's defense system.

2. The U.S. can gain much more by doing business with the Soviet Union than by trying to withhold many things. This latter procedure hasn't been effective where it has been tried.

3. Most important to maintaining a strong international position is that we maintain a strong domestic economy with an adequate number of jobs, especially for young, disadvantaged persons. (If you have studied world-wide Communism you are aware that it preys upon countries where the domestic economy is sick, weak or corrupt).

Further, Mr. Schlafly's letter to you was based on a highly inaccurate booklet by Miles Costick. We are familiar with Mr. Costick's work, and have met with him to document inaccuracies in it. Apparently, he has chosen to ignore this evidence, because:

- While we did sign an agreement for technological cooperation with the Soviet Union, we have **not** transferred any computer technology to them.
- No Control Data computer is being used for any military purpose whatsoever by the Soviet Union.
- Further evidence of error was the blue card that you were asked to mail to Texas Instruments praising them for refusing to sell to the Soviets. I am sure that their chairman, Mr. Bucy, would admit that many Texas Instrument's high technology products are used in Soviet-built equipment.

My position is further explained in the enclosed article, "High Technology Trade with the Communists", which appeared in *Datamation* magazine last January.

I have also enclosed a clipping from a recent issue of *Time* magazine that describes some of Control Data's activities in helping to provide more jobs for the disadvantaged.

If you have questions after reading the material, write me.

Sincerely,

s/s William C. Norris
William C. Norris

Encls.

Letter (Protocol) of Intent dated 19 October 1973 (English version) between State Committee of the USSR Council of Ministers for Science and Technology and the Control Data Corporation.

Letter (Protocol) of Intent

I. The State Committee for Science and Technology of the USSR Council of Ministers and Control Data Corporation (CDC) in the interest of developing extensive long term technological relationships between Soviet organizations and enterprises and Control Data Corporation, have on 19 October 1973 concluded an "Agreement on Scientific and Technical Cooperation". The State Committee for Science and Technology and Control Data Corporation, together with the USSR Ministry of Foreign Trade, hereafter referred to as the Parties, having in mind furthering the development of the desired cooperation, wish to conclude this Letter of Intent to identify joint projects which need detailed study, and to commit the Parties to initiate these studies.

II. The objective of the projects called for in Paragraph III of this Letter (Protocol) of Intent shall be to:

   A. Establish the coordinating group, and working groups called for in Article 4 of the Agreement on Scientific and Technical Cooperation.

   B. Specify the scope and mechanism by which the scientific and technical information exchanges to be engaged in under the terms of Article 2 of the Agreement on Scientific and Technical Cooperation shall be carried out:

   C. Implement detailed examination of those projects given first order of priority.

   D. Define mutually satisfactory commercial and trade relationships.

III. The following preliminary list of tentative joint projects has been agreed to and a detailed examination of them will be our first priority.

   A. **Finance and Trade** — To conduct studies and discussions in order to develop mutually satisfactory bases for payment for the activities to be undertaken within the scope of the Agreement on Scientific and Technical Cooperation. This shall include terms of credit, repayment, considerations, and cooperative activities whose purpose is to generate Western currencies to facilitate repayment of credit.

   B. **Software and Applications Development** — Development by the USSR of automated programming methods for improving programmer productivity as well as the generation of systems within certain applications areas of mutual interest. These would include, but not be limited to, the medical/health care field, the transportation industry, and the education environment.

Objectives of transportation study might include both the increase of efficiency of passenger transportation, as well as enhance the timely delivery of freight, both by surface and by air.

The medical health care study would include the creation of systems dealing with patient care, health care management, as well as health care planning.

Work in the education field will be centered on the disciplines of computer-based education and training, including networks, special terminals, author language, curriculum and courseware.

C. **New Computer Sub-Systems R & D** — The USSR will contribute research and technical development of computer elements based on new technical concepts. This research and development will be undertaken based on cooperative planning between Control Data Corporation and appropriate USSR organizations.

This research and development may include, but will not be limited to, the realm of memory systems, advanced mass memories, thin film, disk heads, plated disks, memory organization and ion beam memories, might be amongst the technologies to be investigated and research and development undertaken.

D. **Reserch and Development** — Joint exploration of research and development activities in USSR enterprises and institutes which will add to Control Data Corporation's research and development activities, particularily in the areas of computer systems software and applications software, and will constitute in part compensation for technologies which the USSR wishes to obtain from Control Data Corporation.

E. **Advanced Computer System Development** — To conduct a cooperative development of a computer with rearrangeable structures and a performance rate of 50 to 100 million instructions per second to be used for solving large economic and management problems. Results of software and architectural research conducted in the Soviet Union will be utilized. The functional design of the software system and the architectural concepts will be supplied by the USSR. The implementation would be conducted in four phases as follows:

1) Feasibility
2) Detailed design
3) Prototype construction and checkout
4) Test and evaluation

The feasibility phase will be used to evaluate the proposed computer concepts, both hardware and software, in order to verify that the system can be implemented with known hardware technology.

The following are among the areas to be considered as part of the evaluation: architectural design, software system design, new software techniques, new software applications, components and circuitry, system performance, operating systems, principal applications, manpower and cost. The entire program is expected to require from five to eight years to reach fruition. Results will be shared on an equal basis.

F. **Disk Manufacturing Plant** — To build a plant for manufacturing mass storage devices based on removable magnetic disk packs with up to 100 million byte capacity per each pack. The yearly plant output shall be 5,000 device units and 60,000 units of magnetic disk packs (approximate estimate). It is expected that 80% of the plant output will be 30 mega-byte devices and 20% will be 100 mega-byte devices.

G. **Printer Manufacturing** — To build a plant to manufacture line printers to operate at a speed of 1,200 lines per minute. The yearly output shall be 3,000 devices (approximate estimate).

H. **Process Control Devices Manufacturing** — To build a plant for the manufacture of process control oriented peripheral devices, including data collection, analog/digital gear, terminals etc. The annual plant output for all devices, including data collection, is estimated as approximately 20,000 units.

Note: With respect to Items F, G and H, the Parties shall jointly develop technological documentation for manufactured products in accordance with the metric system standard and insure that all aspects of the plant(s) design and construction are of the highest quality such that the manufactured output will be fully competitive in all respects within the world market place. Control Data Corporation shall deliver technical documentation, complete sets of technological equipment, as well as know-how, and full assistance in order to insure mastering of full production techniques.

I. **Printed Circuit Board Manufacturing** — Delivery to the USSR of a complete set of equipment for the manufacturing of multilayer printed circuit boards.

J. **RYAD/Control Data Corporation Information Processing Systems** — Joint creation of information processing systems based on the use of Soviet manufactured computers and Control Data Corporation equipment. Control Data Corporation equipment would include local peripheral sub-systems as well as communications hardware, including front-ends, remote concentrators and terminals. It is expected that the system would include operational software jointly created to make a viable computer system for applications to problems in other parts of the world which could be commercially saleable as a total system.

K. **Process Control and Remote Communications Concentrator Manufacture** — To organize within the Soviet Union manufacturing of Control Data Corporation licensed remote communication equipment and analog to digital components for standalone use within technological process control systems. It must be kept in mind that such devices and components must satisfy requirements of both the Soviet Union and Control Data Corporation. Control Data Corporation evaluates that it can buy back approximately $4,000,000 worth of these products.

L. **Network Information System (CYBERNET)** — Joint study to provide a proposed specification for a Soviet computer communication network. Dimensions of networks to be explored will include both the distribution network between terminal users and processing center as well as an eventual bulk transfer network between cluster centers and private satellite centers. The USSR will develop systems and applications software for this Network Information System based upon the present commercially available Control Data Corporation network software system. The resultant software will be available and shared by both Parties. The various data communications technologies will be evaluated as one of the major tasks of the study.

M. **Control Data CYBER 70 Data Processing Centers** — Joint creation of processing centers which will utilize computers and other equipment of Soviet manufacture and those of Control Data Corporation for organization/application such as, but not limited to, those specified below:

1) World Hydrometeorological Institue, Moscow, for processing of weather data for forecasting and in preparation of weather maps.

2) Institute of High Energy Physics, Scrpukhov, for nuclear research, reduction of accelerated data, and basic particle research.

3) Ministry of Geology, Moscow, for seismic data reduction.

4) USSR Academy of Sciences, Academ Goradok, Novosibirsk, for operational numerical weather forecasts, as well as for on-line concurrent processing of scientific data from experiments related to atmospheric, gas dynamics, weather research and basic scientific research.

5) Ministry of Chemical Industry, Moscow, for an information processing network and plant and pipeline control.

6) Ministry of Oil Refining, Moscow, for electric distribution and plant design.

IV. In order to launch the above noted programs into viable, successful implementations, Control Data Corporation is prepared to advance long term credit on mutually beneficial terms through Commercial Credit Company, its financial affiliate. It is expected that the total volume of credit required for financing of the full dimensions of the Scientific and Technological Agreement may reach 500 million dollars. It is further expected that repayment of such credit will be made in the form of Western currencies. Commercial Credit Company is prepared to assist the USSR in obtaining such currencies by working jointly with the appropriate authorities in the USSR in the worldwide marketing of Soviet products and in the development of natural resources for which sustained world demand exists.

In the process of developing specific programs to generate currency for credit repayment, Control Data Corporation offered the following proposals for consideration:

A. **Financing** — Control Data Corporation is prepared to assist the USSR in obtaining the financing needed to accomplish the above noted objectives. Through its wholly-owned subsidiary, Commercial Credit Company, Control Data Corporation can marshal substantial long term credit to facilitate Soviet purchases of equipment and technology projected under the Agreement, as well as credit for potential customers for worldwide sales of Soviet products and materials.

Commercial Credit Company will take the initiative in assembling consortiums of major leaders to extend long term credit to the USSR on mutually beneficial terms. Commercial Credit Company is also prepared to use its Luxembourg facility to make use of U.S. Export/Import Bank financing to the extent such credit is available to finance the export of Control Data Corporation equipment to the USSR.

B. **Joint Venture Marketing Company** — A marketing company jointly owned by a competent Soviet organization and

Commercial Credit Company will be established. The company would be incorporated in Western Europe and would be a self-sufficient, profit oriented entity.

The prime purpose of the company would be to sell the service non-computer-related Soviet products in the Western world on the open market. In addition to selling the products of the USSR the Joint Venture Marketing Company will develop, in conjunction with Commercial Credit Company, economic and trade forecasts for the near to medium term future in the Western world that will indicate market trends, requirements and shortages, which can be fed back into the USSR five-year planning activities to insure that future USSR exports meet the requirements of future Western markets.

C. **Natural Resource Development** — Soviet natural resources are the largest of any single country in the world. The key to world trade in excess natural resources, not required for domestic use, is the creation of additional facilities to obtain and profitably dispose of these resources.

The financial resources and worldwide industry associations of Commercial Credit Company will be employed to create appropriate consortia having four major functional entities:

a. finance
b. resource extraction and processing expertise
c. construction and operation expertise
d. sales outlet for processed products

Natural resources to be considered would include, but not be restricted to: timber products, non-ferrous metals, precious metals and stones, asbestos, apatite, cement, synthetic rubber and resins, and coal.

The present Letter of Intent is executed in two versions, both in Russian and English. Both versions of the text have equal validity.

For the State Committee
of the USSR Council of
Ministers for Science
and Technology

For Control Data
Corporation

For the USSR Ministry
of Foreign Trade

English version of Agreement between State Committee of the Council of Ministers of the USSR for Science and Technology and Control Data Corporation (signed by Robert D. Schmidt), dated 19 October 1973.

On Scientific and Technical Cooperation between the State Committee of the Council of Ministers of the USSR for Science and Technology and Control Data Corporation (USA)

The State Committee of the Council of Ministers of the USSR for Science and Technology (GKNT) and Control Data Corporation (CDC), hereinafter called "Parties",

Considering that favourable conditions have been created for extensive development of a long-term scientific and industrial and economic cooperation;

Taking into account the mutual interest of both Parties in the development of such cooperation and recognizing the mutual advantage thereof; and

In accordance with Paragraph 8 of the "Basic Principles of Relations between the Union of Soviet Socialist Republics and the United States of America", signed on May 29, 1972, and Article 4 of the "Agreement Between the Government of the USSR and the Government of the USA on Cooperation in the Fields of Science and Technology" concluded on May 24, 1972;

Have agreed as follows:

ARTICLE 1

The subject of the present agreement has to do with a long-term program for a broad scientific and technical cooperation in the area computational technology, and specifically;

- To conduct joint development of a technically advanced computer;
- Joint development and organization of the production of computer peripheral equipment;
- Joint creation of information processing systems based on the technical means of Soviet production and on the technical means developed by CDC and the development of software means for these systems;
- Joint development of Analog to Digital Equipment for control systems of technological processes;

—Joint development of computer components, technical equipment for their production and the organization of production of these components.

—Development of computer memories (based on large volume removable magnetic disk packs, and on integrated circuits, etc.).

—Creation of equipment and systems for data communication;

—Application (use) of computers in the fields of medicine, education, meteorological, physics, and etc.;

—Preparation (training) of specialists in the area of computer technology;

The scope of this Agreement may at any time be extended to include other fields of specific subjects of cooperation by agreement of the Parties.

This Agreement is not limiting either Party from entering into similar cooperation in the said fields with a third Party.

ARTICLE 2

Scientific and technical cooperation between the Parties can be implemented in the following forms with specific arrangements being exclusively subject to mutual agreement between appropriate Soviet organizations and the firm of Control Data Corporation:

—Exchange of scientific and technical information, documentation and production samples;

—Exchange of delegations of specialists and trainees;

—Organization of lectures, symposia and demonstrations of the production samples;

—Joint research, development and testing, exchange of research results and experience;

—Mutual consultations for the purpose of discussing and analysing scientific and technical problems, technical principles, ideas and concepts in the appropriate area of cooperation;

—Creation of temporary joint research groups to perform specific projects and to produce appropriate (joint) reports.

—Exchange, acquisition or transfer of methods, processes, technical equipment, as well as of "know-how" and of licenses for the manufacture of products.

ARTICLE 3

The Parties have established that financial, commercial, and legal questions related to advancement of credit and payments for the delivered products and technical equipment, assignation of licences and "know-how" as well as supplied services in performance of the various joint projects, relative to the present Agreement, shall be decided by separate agreements between appropriate competent Soviet organizations and the Control Data Corporation.

ARTICLE 4

For the practical implementation of the present Agreement the Parties shall establish a Coordinating group, from authorized representatives (coordinators) which shall determine and recommend a proper course for the cooperation and also to control compliance with responsibilities assumed by the Parties, and to take the necessary action for the successful implementation of the objectives of the present Agreement. For the preparation of proposals for the concrete cooperative projects, there shall be established special groups of experts whose task it will be to determine technical and economic feasibility of the joint projects and to draw up action plans for their realization. The results of these working groups shall be turned over to the Coordinating group for their discussion and preparation of recommendations.

Recommendations and proposals of the Coordinating group will be presented in the form of protocols, which will be used as the basis for preparation of separate protocols or contracts.

Coordinating and working groups shall meet as frequently as is necessary to perform their functions alternativelly in the USSR and USA unless otherwise agreed.

ARTICLE 5

Scientific and technical information furnished by one Party to the other under this agreement may be used freely for its own research, development and production, as well as the realization of finished products unless the Party supplying such information stipulates at the time of its transfer that the information may be used only on the basis of special agreement between Parties. This information can be transmitted to a third Party only with the approval of the Party which has furnished it.

Information received from a third Party which cannot be disposed of at will by one of the Contracting Parties is not subject to transmittal to the other Party unless mutually satisfactory arrangements can be made with the third Party for communication of such information.

It is contemplated in the foregoing that any organizations or enterprises of the USSR and any wholly owned or partially owned Control Data subsidiaries shall be not regarded as a third Party.

ARTICLE 6

Expenses of travelling back and forth of specialists of both Parties under the programs related to this Agreement, as a rule will be defrayed as follows:

—The Party sending the specialists pays the round-trip fare.
—The host Party bears all costs connected with their stay while in its own country.

The duration of the above visits and the number of specialists in each group shall be mutually agreed to by the Parties in advance of the visits.

Organizational questions, arizing from implementation of this present Agreement shall be discussed and determined by the Parties in the course of working.

The present Agreement shall continue for a period of 10 (ten) years and shall enter into force immediately upon its signature. It can be extended with mutial agreement of the Parties.

The cancellation of the present Agreement shall not affect the validity of any agreement and contracts enetered into in accordance with Article 3 of the present Agreement by organizations and enterprizes of the USSR and CDC.

Drawn up and signed the 19 October 1973, in the city of Moscow, USSR, in duplicate, one copy in Russian and one in English, both texts being equally authentic.

For the State Committee of the
Council of Ministers of the USSR
for Science and Technology
/s/ (illegible)

For the Control Data
Corporation
/s/ Robert D. Schmidt

Appendix E:

Position of Texas Instruments Company and Chairman Fred Bucy on dangers of trading technology to the Soviets. (This letter demonstrates that the identification of Deaf Mute Blindmen must be made with care)

TI's Fred Bucy Warns Against Selling Technology Know-How and Turnkey Semiconductor Plants to Communist Nations

Even as West and East Bloc countries discuss relaxing the trade embargo on high technology, a sharp disagreement has developed between two major semiconductor manufacturers on how far this relaxation should extend. One viewpoint is exemplified in last month's Wescon speech by C. Lester Hogan, president of Fairchild Camera & Instrument Corp., who wants few restrictions made in East-West semiconductor trade.

Hogan stands ready to sell semiconductor products as well as the production equipment, technology, and management know-how to the Communist countries. This would mean the Communists would manufacture high-level semiconductor products themselves. The French company, Sescosem, a division of Thomson-CSF, has already broken the ice with the first turnkey semiconductor plant — sold to Poland — and Fairchild, waiting for the trade detente, does not deny having one quote outstanding on an MOS plant for Poland, and another two behind that for Russia.

On the opposite side is Texas Instruments, whose executives take a dim view indeed of trade agreements with the Communist bloc, on the grounds that such agreements are not protected by patent rights and do not offer either open markets or the opportunity to build a decent market share. J. Fred Bucy, vice president of Texas Instruments, points out that "it's one thing to sell high-technology products in the foreign market, but quite another to sell the know-how to make these products." He is adamant against turnkey contracts, and sees an equal risk in selling the Communists such pieces of production equipment as line-and-expose towers, diffusion furnaces, epitaxial reactors, and the like. "It's axiomatic in high-technology industries," says Bucy, "that the only adequate payment for know-how is market share. No lump-sum payment or turnkey-service fee can be great enough to fund the research and development necessary to enable the seller to maintain his advantage. You can be sure," he emphasizes, "that if we give away the know-how without obtaining a market share, they won't buy a dime from us — devices or equipment. We will be giving away the crown jewels."

In Bucy's eyes, a big question with East-West trade is patent recognition. Because of the longstanding embargo on high-technology trade, few Western semiconductor patents are recognized by the Russians or other Eastern Europeans. "Since we can sell into the Communist community only through their governments," says Bucy, "the only way we can participate in their markets is for them to agree to recognize our patents retroactively, pay us full royalties, and/or give us access to sell directly in their markets on an equal footing against their state-owned factories."

Bucy also feels "that if the free world is footloose with its technology, it may be building a monster that soon will gobble up domestic markets. The Communist countries will build their semiconductor capability with Western-supplied production equipment and know-how, and/or with Western-built turnkey plants protected by high tariff barriers," he says, and when they get their costs down and their own markets saturated, they will be "right out there exporting into ours."

This means that any semiconductor plant set up in Eastern European countries capable of anywhere near the capacity of typical Western manufacturing facilities will quickly saturate the domestic Communist market and be ready to export their overcapacity to the West.

The semiconductor capability of the Comecon has been greatly exaggerated as well, according to Bucy. "The shell game they're playing," says Bucy, "is taking small quantities of laboratory-developed devices, giving them to people who are visiting Russia, and saying 'look at our capability — now why not sell us the equipment to manufacture this, because we can do it ourselves anyway.' If they can, let them do it. The truth is, they don't have the capability of producing in large quantities at high yields. And that's what they want us for."

Appendix F:

U.S. Firms Trading with the Soviet Union in the 1960-1985 Period

Data concerning firms trading with the Soviet Union and the technologies transferred is classified by the Department of Commerce, allegedly to protect business from competitors. Censorship also has the effect of preventing independent analysis and public opinion from coming to grips with Soviet trade.

The following list was compiled from official files for the early and mid 1970s, supplemented by corporate news reports for the later years.

It is incomplete but certainly includes all major U.S. operations in the Soviet Union, and is the only such list in existence in the public domain.

Company	Product
AEC	Atomic Energy
Acme Mfg. Co.	Machine Tools
Alcoa	Aluminum
Allen Bradley	Machine Tools
Alliance Tool & Die Corp.	Machine Tools
Alliance Tool & Die	Machine Tools
Allis-Chalmers	Machine Tools
Allsteel Press Co.	Machine Tools
Alpha Press Co.	Non Ferrous Metals
American Can Co.	Iron & Steel
American Can Co.	Food Machinery
American Chain & Cable	Machine Tools
American Express	Services
American Magnesium Co.	Metals Technology
Applied Magnetic Corp.	Electronics
Ara Oztemal (Subsidiary of Satra Corp.)	Trading
Armco Steel	Steel
Atlas Fabricators Inc.	Machine Tools
Automatic Production Sys. (Div. of Ingersoll-Rand)	Motor Vehicles
Babcock & Wilcox	Boiler Technology
Bechtel	Construction
Belarus Equip. of Canada Ltd.	Agriculture Equip.
Bendix Corp.	Machine Tools
Besley Grinder Co.	Machine Tools
Bliss, E.W. Div. of Gulf & Western Industries	Motor Vehicles
Boeing	Machine Tools
Boeing	Aircraft Technology
Borg-Warner	Machine Tools

Brown & Root Inc. . . . . Construction
Brown & Sharpe Mfg. Co. . . . . Machine Tools
Brunswick Corp. . . . . Machine Tools
Bryant Chucking Grinder Corp. . . . . Machine Tools
Burr-Brown Research Corp. . . . . N.A.
C-E Cast Equipment . . . . Machine Tools
Carborundum Co. . . . . Motor Vehicles
Carlton Machine Tool Co. . . . . Machine Tools
Carpenter Technology Corp. . . . . N.A.
Caterpillar Tractor Co. . . . . Agriculture Equip.
Centrispray . . . . Machine Tools
Century Data . . . . Electronics
*Chase Manhattan Bank . . . . Finance
Chemetron Corp. . . . . Machine Tools
Cincinnati Milacron Inc. . . . . Machine Tools
Clark Equipment . . . . Machine Tools
Cleveland Crane & Eng. . . . . Machine Tools
Colonial Broach . . . . Machine Tools
Combustion Engineering . . . . Motor Vehicles
Comma Corp. . . . . Electronics
*Control Data . . . . Advanced Computers
Cooper Industries Inc. . . . . Petroleum Equipment
Cromalloy-Kessler Asso. Inc. . . . . Machine Tools
Cross Co. . . . . Machine Tools
D.A.B. Industries Inc. . . . . Prime Movers
Denison Div. of Abex Corp. . . . . Machine Tools
DoAll Co. . . . . Machine Tools
Douglas Aircraft . . . . Machine Tools
Dow Chemicals . . . . Chemicals
Dresser Industries . . . . Oil Tool Equip.
Dr. Dvorkovitz & Asso. . . . . Non Ferrous Metals
E.I. duPont de Nemour & Co. . . . . Chemicals
E.W. Bliss, Div. of Gulf . . . . Machine Tools
Easco Sparcatron . . . . Machine Tools
Electronic Memories & Magnetics Corp. . . . . . Electronics
El Paso Natural Gas Co. . . . . Gas Technology
Englehard Minerals & Chem. Corp. . . . . Machine Tools
Ex-Cell-O Corp. . . . . Machine Tools
FMC . . . . Machine Tools
Fenn Rolling Mills Co. . . . . Rolling Equip.
Fon du Lac . . . . Machine Tools
Ford Motor Co. . . . . Non Ferrous Metals
GMC . . . . Machine Tools
Gearhart-Owen . . . . Petroleum
General Dynamics . . . . Aeronautical Tech.

General Electric......................Petroleum Equip.
General Electric......................Machine Tools
General Tool Corp.....................Machine Tools
Giddings & Lewis ....Machine Tools
Gleason Works ....Motor Vehicles
* Gleason Works ....Machine Tools
Goddard Space Flight Center ....Machine Tools
Gould Inc. ....Motor Vehicles
Gulf General Atomic ....Atomic Energy
Gulf Oil Corp.........................Petroleum Equip.
Halcroft & Co. ....Machine Tools
Harig Products, Inc. ....Machine Tools
Hewlett-Packard ....Electronics
Holcroft & Co. ....Motor Vehicle
Honeywell ....Computers
Honeywell Information Systems ....Electronics
Hudson Vibratory Co...................Mechanical Equip.
IBM ....Machine Tools
IBM ....Computers
IBM ....Electronics
Ingersoll ....Machine Tools
Ingersoll Milling Machine ....Motor Vehicles
Ingersoll Rand Co. ....Machine Tools
* International Computers ....Electronics
Intel Corp. ....Electronics
International Nickel ....Nickel Technology
International Harvester ....Machine Tools
Irving Trust Co.......................Finance
Itel ....Electronics
Jones & Lamson (Textron) ....Machine Tools
Joy Mfg..............................Drilling Equip.
Kaiser Aluminum ....Non Ferrous Metals
Kaiser Aluminum & Chem. Corp. ..........N.A.
Kaiser Industries Corp. ....Machine Tools
Kearney & Trecker Corp. ....Machine Tools
Kennametal Inc. ....Machine Tools
Kingsbury Machine Tool Corp. ....Machine Tools
Koehring Co. ....Machine Tools
LaSalle Machine Tool Co...............Machine Tools
Leasco ....Electronics
Leesona Corp. ....Textile Equip.
Libby-Owens-Ford Co. .................N.A.
Litton Indust. Int'l. Mach. Tool Syst. Group ...Machine Tools
Lockheed ....Aircraft Technology
Lummus Corp..........................Oil Technology

McDonnell Douglas . . . . Aircraft
Micromatic Hone . . . . Machine Tools
Minnie Punch & Die Co. . . . . Machine Tools
Modicon Corp. . . . . Machine Tools
Monarch . . . . Machine Tools
Monsanto . . . . Chemical
Moore Special Tool Co. . . . . Machine Tools
NBS . . . . Machine Tools
National Engineering . . . . Machine Tools
NED's Reactor & Fuel Mfg. Fac. . . . . Atomic Energy
Norton Co. . . . . Grinding Equip.
• Occidental Petroleum . . . . Major factor in
                                                      complete plants
Pneumatic Tool Co. Machine Tools
Pratt & Whitney . . . . Machine Tools
Pullman Corp. . . . . Machine Tools
RCA . . . . Electronics
Raycon Corp. . . . . Machine Tools
Raytheon . . . . Electronics
Reynolds Metals Co. . . . . Metals
Rockwell International . . . . Machine Tools
Sikorsky Aircraft . . . . Aircraft
Singer Co. . . . . Machine Tools
Snow Mfg. Co. . . . . Machine Tools
Sorbus Inc. . . . . Electronics
Spectra-Physics Inc. . . . . Electronics
Speedfam Corp. . . . . Machine Tools
Sperry Rand . . . . Machine Tools
Standard Oil . . . . Oil Technology
Sundstrand Corp. . . . . Machine Tools
Swindell-Dressler Co. . . . . Motor Vehicle
Swindell-Dressler Co. . . . . Machine Tools
Systrom-Donner Corp. . . . . Electronics
Technic Inc. . . . . Non Ferrous Metals
Tektronix Inc. . . . . Electronics
Teledyne Landis . . . . Machine Tools
Teledyne Pines Co. . . . . Prime Movers
Tenneco . . . . Chemicals
Texas Eastern Transmission . . . . N.A.
Textron Inc. . . . . Machine Tools
Udylite Corp. . . . . Machine Tools
Univac . . . . Electronics
V & O Press . . . . Machine Tools
VSI Automational Assembly Inc. . . . . Machine Tools
Varian . . . . Electronics

*These are the key companies involved.

Appendix G:

Confidential Government Report on Cummins Engine Company (J. Irwin Miller) and Financing of Marxist Revolutionary Activities Within the United States.

I. INTRODUCTION

This Section has been reviewing available data on various philanthropic foundations in an effort to determine if their activities are consistent with their announced goals. In this instance we have made a preliminary analysis of the Cummins Engine Foundation which although a legitimate philanthropic enterprise, nonetheless would appear to serve as a conduit for funds to support black militants and organizations which are known to consistently promote radical revolutionary activities.

This analysis roughly outlines the data presently available. Obviously, there is considerably more to be learned concerning CEF's overall activities. Additional inquiries to the FBI are contemplated and liaison has already been established with the Internal Revenue Service in an effort to develop our respective interests.

Unfortunately, the complex nature of this *tax exempt* foundation does not lend itself to an easy explanation. It is suggested that the reader make frequent reference to the attached chart which will hopefully enhance understanding of the organization.

II. BACKGROUND

The Cummins Engine Foundation (CEF) was established in 1954 as a philanthropic organization by J. Irwin Miller, Chairman of the Board, Cummins Engine Company, Columbus, Indiana. Mr. Miller is a highly respected 62-year old civic leader who in recent years has devoted much of his time and money to various charities and religious pursuits. He is regarded as a part of the "avant garde of the civil rights movement."

CEF is financed solely by Cummins Engine Company, which is the parent of a worldwide concern reputed to be the largest independent producer of high speed diesel engines. In 1970 CEC's pretax profit was $44,564,365.00, of which five percent was allocated to CEF. CEF generally receives one to three million dollars annually.

CEF is comprised of two sections: (1) the original County Related Philanthropic (CRP), which provides money for educational, health and social welfare projects in Bartholomew County, Indiana and (2) the General Philanthropic (GP), which was established in 1968. Little is known about the operations of the former and this report deals exclusively with the known activities of the GP section, which shares the pretax profits with the CRP.

In an effort to assist in the cultural and economic development of the nation's black people, CEF's GP section provides large sums of tax-free dollars which are distributed through program directors in five cities, viz., Baltimore, Washington, D.C., Los Angeles, Atlanta and Chicago. There appear to be no auditing procedures once the funds have been allocated by CEF to the program directors. This situation creates broad opportunities for misdirection or misappropriation of monies. The overall thrust of this report, therefore, is to illustrate how a seemingly munificent activity can, by divers means, serve to aid the objectives of radical individuals and organizations.

Specifically, CEF has provided a vehicle whereby:

(1) Monies were given to one organization earmarked for the use of another organization which in turn has recruited black activists to participate in revolutionary training in Communist China in December 1971:

(2) CEF contributed $5,500.00 during 1970 to the Institute for Policy Studies (IPS). A Resident Fellow of IPS also serves as the Washington Program Director for CEF;

(3) CEF funds are utilized for the purchase and operation of printing equipment ultimately used by groups advocating revolutionary social change;

(4) A close working relationship exists between organizations funded by CEF and certain Marxist-Leninist organizations. One of these is headed by James Forman, a principal advocate of black revolutionary training abroad;

(5) A CEF Program Director was indicted for inciting to riot during a demonstration at a penitentiary in October 1970 while serving as a self-appointed spokesman for prisoners' grievances:

(6) The aforementioned Program Director has purchased a 130-acre farm in his name, apparently using CEF funds. The announced purpose for the farm is that it is to be used as a youth camp to "train children in self-defense and in African culture";

(7) Seemingly irregular fiscal practices exist with respect to CEF-funded organizations in Baltimore and inordinate amounts of money are dispensed for such items as "materials";

(8) The possibility exists that certain "dummy" organizations have been created in order to siphon-off funds which are destined for otherwise legitimate activity.

III. DETAILS

To date the primary area of investigation into CEF activities has been concentrated in Baltimore, Maryland. The paucity of information concerning CEF funded activities in Los Angeles, Atlanta, Chicago and New York stems from the fact that investigation has not yet been actively instituted in those areas. For this reason the scope of this report will be limited mainly to those activities of CEF in the Baltimore, Maryland area.

The CEF Program Director in Baltimore is Walter Lively, a 29-year old black activist who since 1961 has been variously involved with the Socialist Party, CORE, U-JOIN (a jobs for blacks program he organized in Baltimore), SNCC and the Urban Coalition. He now heads the Baltimore Neighborhood Commons (BNC), a corporation through whose bank account CEF channels a portion of its money. Presumably, CEF money enabled Lively to establish the Liberation House Press, which currently offers a full range of printing services at cut-rates to "radical groups." Lively, with his fellow BNC officers, Austin Sydnor and Kenya Kiongozi, also operates the Liberation Bookstore. The BNC, the Liberation House Press, Liberation Bookstore, and an office of the International Black Workers Congress (IBWC) are all located at the same address in Baltimore. Lively, Sydnor and Kiongozi also maintain close liaison with the IBWC, the Black Panther Party (BPP) and SOUL School (SS), and significantly, these individuals were arrested and indicted for inciting to riot at a BPP-SS rally at the Maryland State Penitentiary in October 1970.

In December 1970, Lively purchased for $18,000.00 a 130-acre farm in Bedford, Pennsylvania, where Liberation Press and SOUL School are to build a youth camp to "train children in self-defense and African culture." (SOUL School is a "non-profit educational institution" whose literature in the past has advocated black revolution.)

The manner in which BNC funds are expended seems to suggest a lack of auditing procedures. From December 21, 1970 to September 17, 1971 a total of $46,887.62 was withdrawn from the BNC account. Of this amount, checks totaling $17,135.00 were made payable to Cash or an officer of the BNC, usually for reason of "materials" or "supplies." It should be noted that checks totaling $2,-830.14 were made payable to companies which deal in paper and printing supplies and $11,508.00 was withdrawn for purchase of a printing press. Similarly suspect are the activities of organizations which have received CEF funds through BNC and Lively.

In 1971 CEF made available $12,000.00 to BNC designated for the National Association of Black Students (NABS), an organization whose stated purpose is to provide a communications network for black college students. As of November 1971 NABS was recruiting revolutionary black students to travel to China with a group recruited by William Epton's Marxist-Leninist Collective. The express purpose of the trip is to provide training for student organizers to further revolution in the United States.

At least $5,000.00 from CEF was sent to BNC in 1971 marked for the Frantz Fanon Institute (FFI). FFI is allegedly an educational entity dedicated to inculcating Marxist-Leninist teachings in the minds of workers and students. Although a reference to its inception was made by radical activist James Forman at a NABS meeting in December 1970, the actual location of FFI is presently unknown. The available description of FFI's structure and program was found attached to an International Black Workers Congress (IBWC) mailing list. It should be noted that James Forman founded IBWC and traveled to Algeria in 1970 ostensibly to gather information for a book he was writing about Frantz Fanon.

In September 1971 the three officers of the BNC traveled to Gary, Indiana for the IBWC conference called by James Forman. The purpose of the conference was to instruct local chapters in the techniques of organizing workers in factories to further the cause of destroying capitalism and building a socialist state.

Forman, who traveled to Hanoi and Peking in July 1971, had copies of the 16 page *Manifesto of the IBWC* distributed to the 200 conferees, who were present by invitation only. This document includes, among other objectives: the elimination of "parasitic capitalistic vultures"; total disarmament of the U.S.; destruction of all police forces and their replacement with peoples militia; and an end to the blockade of Cuba. The program designed to effect these goals stresses teaching workers and students Marxist-Leninist ideology and promotes acceptance of money from white institutions as a good revolutionary tactic. The *Manifesto* concludes with a 6-point methodology for revolutionary work quoted from a 1970 North Vietnamese publication. The precise nature of the relationship betwen IBWC and BNC is unclear. It is nonetheless noteworthy that in April 1971 CEF'S program directors were said to be attempting to organize black American workers.

Ivanhoe Donaldson, CEF's Washington, D.C. Program Director, also serves as a Resident Fellow for the Institute for Policy Studies (IPS) which has been characterized as "a radical think tank." He was selected for the CEF position because of his contacts with "leading black activists." Donaldson served concurrently as a leader for the Student Non-Violent Coordinating Committee (SNCC) in New York City and as an IPS fellow.

During the period 1967–1969 Donaldson associated with Stokely Carmichael and SNCC leaders Stanley Wise and James Paul Garrett. In October 1969, Garrett opened the Center for Black Education (CBE) which in 1971 received a lump sum of $10,000.00 from IPS. CBE is characterized as an educational institution independent of and opposed to the aims of the American nation, and dedicated to the liberation of African people.

CBE affiliates are involved with the African-oriented Drum and Spear Bookstore, which was founded by Ivanhoe Donaldson and is managed by Marvin Holloway, who is also associated with IPS. Two other IPS affiliates who are on the CBE staff operate the Drum and Spear Press in Washington, D.C. Interestingly, CEF's list of projects includes a "Drummonds Sphere Press" in New York, whose purported function is to print black children's literature. To date no information has been developed to verify the existence of "Drummonds Sphere," whose semantic similarity to Drum and Spear would seem more than coincidental.

A survey of organizations involved in the CEF financial web reveals consistent efforts to disseminate revolutionary materials and information. For example, IBWC's program calls for the establishment of bookstores and printing concerns to better spread the word on Marxism-Leninism. Gwen Patton, founder of NABS, stated at an IBWC conference that efforts are being made to establish bookstores on or near black campuses and communities to circulate "political education" material advocating the Marxist-Leninist line. The Liberation House Press in Baltimore, which is owned by Walter Lively, is described as a print shop for "movement groups" and operates in conjunction with the Liberation Bookstore. CBE in Washington, with its close ties to the Drum and Spear Bookstore and Drum and Spear Press, has sent its members to Baltimore to learn to operate Walter Lively's printing press. Still unaccounted for is the mysterious Drummonds Sphere Press in New York.

IV. CONCLUSION

As previously noted the primary investigative activity in this case has been concentrated in the Baltimore, Maryland area. A concerted effort will be made to seek new data and refine that which already exists. It would appear that an exhaustive analysis of the complicated maze of subsidiary relationship and interlocking interests will take on greater significance when examined in detail.

Obviously, additional investigation is required in order to supply information concerning the many unanswered questions, a few of which are as follows:

(1) What are the motives of J. Irwin Miller and the two CEF Directors (James A. Joseph and Roger P. Hoffman) and are these individuals aware of the ultimate, specific use of CEF funds?

(2) How much money goes to the respective programs and program directors?

(3) What is the specific nature and organization of the vehicle used to disburse CEF funds in Atlanta, New York, Washington, D.C., Los Angeles and Chicago?

(4) What is the true relationship between CEF and IPS and what roles do IPS employees Ivanhoe Donaldson, Courtland Cox, Charles Earl Cobb and James P. Garrett play?

(5) What is the nature of the relationship between Center for Black Education and IPS? the Drum and Spear Bookstore? the Drum and Spear Press?

(6) What are the complete circumstances surrounding Lively's purchase of farm land in Pennsylvania and the contemplated use of this land?

(7) What is the relationship between NABS and BNC and what role, if any, does Burnett Pointdexter play?

(8) Does BNC serve any purpose, other than to disburse CEF funds?

(9) Who is Fedha Inya, one of the individuals authorized to sign BNC checks?

In the course of our liaison with IRS, information has been exchanged regarding the activities of individuals and organizations associated with CEF. [Deleted here are references to action the IRS has taken regarding this organization.]

*   *   *   *   *   *   *

Appendix H:

From the Phoenix Letter, January 1986 Issue (Research Publication, P. O. Box 39850, Phoenix, Arizona 85069)

Secret U.S.-Soviet Group Plots Treason

A **secret U.S.-Soviet operating group** based in New York is acting in a manner strongly suggesting treason by its American members — and we do not use the word treason lightly.

We have circulated certain information in our possession to close associates. Without exception they are shocked, shaken, disbelieving. . . **and angry.**

Before we spell out the details for you, how do we define **treason?** Treason is defined in the Constitution as "aid and comfort" to an enemy of the United States.

Second, following this how do we define **an enemy?** We spend $300 billion a year on defense against the Soviet Union, so logically the Soviet Union is an enemy. We cannot SIMULTANEOUSLY have a $300 billion defense budget against the Soviet Union **and** give "aid and comfort" for the Soviet military structure without raising the question of treason.

This topic is discussed at length in a forthcoming book, *The Best Enemy Money Can Buy.*

Who is this secret group? Why do we suggest treason?

US-USSR Trade And Economic Council, Inc. (TEC, for short)

This bland official sounding title **disguises a formal joint Soviet-American apparatus to conduit advanced technology with pure military applications to the Soviet Union.**

The information that follows raises a distinct possibility that Administration "spy trials" and pious proclamations against aiding the Soviets are so much hogwash. . . **a smokescreen.**

This TEC group has official sanction and backing. It links to Bush elements in the White House, the National Association of Manufacturers, the U.S. Chamber of Commerce, and assorted Senators and Congressmen with more political ambition than common sense.

THE US-USSR TRADE AND ECONOMIC COUNCIL, INC. IS A SECRET ORGANIZATION.

TEC WILL NOT RELEASE ITS MEMBERSHIP LIST (ABOUT 300) TO THE MEDIA. TEC WILL NOT EVEN RELEASE ITS LIST TO U.S. FIRMS UNLESS THE FIRM HAS PASSED AN INTERVIEW BY A SOVIET NATIONAL WHOM WE SUSPECT HAS KGB LINKS.

Now hear this:
- The Soviet Government has a list of members
- The U.S. Government has a list of members
- The U.S. PUBLIC IS DENIED THIS INFORMATION, even though TEC claims official backing.

We acquired a **partial** list from a confidential source (reprinted on page 245). We know this partial list is accurate. We also know 500 copies of the membership list exist — so it's only a matter of time and persistence until we access the full list.

TEC claims backing from Vice President Bush (Skull & Bones) and Commerce Secretary Malcolm Baldrige, Jr. (son of S&B member Malcolm Baldrige). This makes TEC a quasi-public institution. Therefore, the public has a RIGHT TO DEMAND THE MEMBERSHIP LIST.

Why all the secrecy? Read on.

What We Know About TEC:
- TEC is headed by:
  VLADIMIR N. SUSHKOV (Co-Chairman) Soviet citizen
  DWAYNE O. ANDREUS (Co-Chairman) U.S. citizen
  JAMES H. GIFFEN (President)   U.S. citizen
  YURI V. LEGEEV (Vice-President)   Soviet citizen
- The PERMANENT directors of TEC are:
  ALEXANDER TROWBRIDGE, President National Association
  of Manufacturers
  RICHARD LESHER, President U.S. Chamber of Commerce
- TEC has 30 Soviet directors and 20 U.S. directors.
- TEC has eight full time Soviet engineers living in New York to assess U.S. technology. The cost is borne by TEC members. THESE SOVIET ENGINEERS INTERVIEW U.S. BUSINESS-MEN AND ASSESS THE U.S. TECHNOLOGY OFFERED. THEY SELECT TECHNOLOGY REQUIRED FOR SOVIET MILITARY END USES AND FACILITATE THE TRANSFER.
- In a recent White House meeting the President's Science Adviser, Dr. George Keyworth, made the following statement:
  ". . . we all know that the Soviets are robbing us blind."
- Furthermore, the White House has a list of **more than 150 Soviet weapons systems using U.S. technology.**
- Yet Vice President Bush and Commerce Secretary Malcolm Baldrige, Jr. are **backing this Soviet technical vacuum operation.** We reproduce on page 246 a copy of a letter sent by Baldrige to many U.S. firms in June 1985. (The marginal notations are by our confidential source. We left them intact.)

Partial List of Membership
U.S.-U.S.S.R. Trade & Economic Council Inc.

Abbott Laboratories
Allen Bradley
Alliance Tool Corporation
Allied Analytical Systems
Allis Chalmers
American Cyanamid
American Express
Archer Daniels Midland
Armco Steel
Bunge Corporation
Cargill
Caterpillar
Chase Manhattan
Chemical Bank
Clark Equipment
Coca Cola
Con Agra
Continental Grain
Corning Glass
Deer & Company
Dow Chemical
Dresser Industries
E. I. DuPont
FMC Corporation

Garnac Grain
Gleason Corporation
Hope Industries
Ingersoll Rand
International Harvester
Kodak
Marine Midland
Millipore Inc.
Minnesota Mining
Monsanto
Occidental Petroleum
Owens Illinois
Pepsi Co.
Phibro-Salomon
Phillip Morris
Ralston Purina
Rohm & Haas
Seagram
Stauffer Chemical
Tendler-Beretz Associates
Tenneco Inc.
Union Carbide
Unit Rig & Equipment Company
Xerox

Alfred J. Murrer, The Key U.S. Operative

The key American operative in the New York office of TEC is ALFRED J. MURRER, former Chairman of the Gleason Works, a large machine tool firm in Rochester, New York with plants in Belgium and West Germany.

Murrer works closely day to day with the Soviet engineers. For our example here we select **Alexander Y. Markov,** listed as a Project Engineer. These two men jointly screen American business companies and if they find the offered technology interesting, attempt to draw the firm into the TEC net.

We can dismiss Alexander Markov briefly. Markov is a **Soviet citizen,** chemical engineer, suspected KGB, resident in New York for almost three years. Markov's job is to identify the usefulness of technology to the Soviet military complex. As Markov is a Soviet national, the question of treason does not arise. So far as we know Markov is a patriotic Soviet citizen doing his job. His only defect (that we

THE SECRETARY OF COMMERCE
Washington, D.C. 20230

A Message To The
American Business Community

T he President is seeking to build a more constructive working relationship
with the Soviet Union, and he favors an expansion of peaceful trade with
the U.S.S.R. as part of this effort. After a seven-year hiatus, the United
States agreed to a Joint Commercial Commission meeting this year.

I recently returned from Moscow, where along with Soviet Foreign Trade
Minister Patolichev, I co-chaired the Eighth Session of the Joint U.S.-U.S.S.R.
Commercial Commission. The Commission, which was established in 1972 to
help expand two-way trade and economic relations, has not met since 1978.

At the Commission meeting, the Soviet government agreed to take steps
which will improve the access for U.S. firms to the U.S.S.R. market. Trade
Minister Patolichev will write to all Soviet Foreign Trade Organizations to in-
form them of the Soviet government's desire to:

● see commercial cooperation with the United States increase by providing
bid inquiries to interested U.S. firms;

● consider U.S. company proposals fully on their economic merits;

● provide U.S. firms with access to Soviet trade and purchasing officials; and

● give them the Agreed Report of the Commission, which goes into more
detail.

I announced at the Commission meeting that upon my return to Washington I
would provide the U.S. business community with a copy of the Agreed Report,
and would encourage U.S. business to explore trading opportunities in the
U.S.S.R. The Agreed Report follows my letter, and I urge you to read it.

The Soviet market is never an easy one, but I believe that U.S. firms trying to
sell in the U.S.S.R. will find the business climate there improved. Let me re-
mind all U.S. companies that any products or know-how to be exported to the
U.S.S.R. must be consistent with our export control regulations. The Department of Commerce is ready to assist firms in complying with these regulations
as well as aiding with their marketing efforts.

Malcolm Baldrige

Secretary of Commerce

have identified) is a habit of chainsmoking Turkish type oval cigarettes which overwhelms American visitors (unused to the pungent odor) in clouds of smoke.

Alfred Murrer deserves more attention because he is a **U.S. citizen,** born Rochester, New York 1923, educated MIT and University of Rochester. Before joining TEC, Murrer was Chairman of the Board of the Gleason Works, with 42 years service.

The Gleason Works is a long established prescision tool maker specializing in precision machine tools for cutting and grinding straight, zerol, spiral, bevel, hyphoid gears and curvic couplings with associated tooling and equipment.

It is vital to note that Gleason Works is in bad financial shape, with heavy losses during the past few years.

Every last machine tool produced by Gleason Works has military applications. We know from experience that there is no purely civilian industry in the Soviet Union. All plants are first and foremost military plants. All technology is examiend first and foremost for military applications. This was demonstrated 20 years ago in my three volume WESTERN TECHNOLOGY AND SOVIET ECONOMIC DEVELOPMENT and since confirmed by dozens of defecting Soviet engineers. This thesis is accepted today by Department of Defense and Central Intelligence Agency. (It is not accepted by Department of State, which operates in a fantasy world of its own making.)

What is the relationship between the financially troubled Gleason Works and the Soviet Union TODAY?

At end 1985 Gleason had a gear-cutting machine population in the Soviet Union in excess of 2500 units, including over 300 at the Kama River Truck Plant, which produces military vehicles for Afghan genocide.

Gleason has trained more than 300 Soviet engineers (actually technicians, but the Soviets classify them as engineers) to maintain these Gleason machines. Their end use is preeminently military.

Gleason services this 2500 machine tool population out of its Rochester, New York and Badour, Belgium plants. In addition, Gleason is building a gear arbor plant for the Soviets (how this ever passed Export Control is a significant unanswered question).

Gleason's subsidiary, Alliance Tool, supplies equipment and systems for manufacture of carbide tooling, vital for many military applications.

Murrer Uses Deceit To Entice U.S. Firms Into The TEC Net

American businessmen faced by **a Soviet national** would presumably be cautious about disclosing technology of military significance. At least one would **hope** that is the case. However, an

American faced by **another American** would assume that his guard can be dropped because the American negotiator is looking out for U.S. national interests.

In the case of TEC we have evidence that American nationals in the organization are working for the Soviet side. Specifically we have evidence that Murrer is using **deceit**, i.e., false statements, to entice U.S. businessmen to transfer technology to the Soviet Union.

Alfred Murrer made the following statement to an American businessman (whom we shall call "X") in the presence of Alexander Markov. Businessman "X" was concerned that his technology might be copied by the Soviets. Not unreasonable, because every New York taxi driver and Chicago bartender knows the Soviets copy our technology.

When businessman "X" queried Murrer about this possibility, Murrer made the following reply:

"No. The Japanese went over a new Gleason cutter with rulers and cameras 'after hours' at a 1965 NMTBA Show (National Machine Tool Builders' Association) to duplicate it. **The Soviets would not do this.**"

This is a demonstrably untrue statement. It is, of course, common knowledge that the Soviets operate copying centers to duplicate the most useful of Western technology. **What is highly significant is that Gleason gear-cutting machines have been duplicated by the Soviets without permission. As Murrer worked 42 years for Gleason and ended up as Chairman of the Board, Murrer knows this.**

Here is an extract from a State Department report (Decimal File 861.797/37):

"Engineers sent by firms were offered individual contracts to stay and work directly for Soviet organizations in order to copy foreign equipment. J. Ubanik, at the (former) Stalin Auto Plant, reported that after copying several of the old Gleason machines, the Soviets tried to enlist his assistance in copying a new Gleason gearmaking machine."

Why would Murrer attempt to deceive a fellow American in favor of the Soviet Union? Particularly when it involves technology with military end use?

If any reader cares to argue this point with Alfred Murrer personally, his private telephone number at the US-USSR Trade and Economic Council is: (212) 644-4568. If they change it, try the general offices of US-USSR Trade and Economic Council at (212) 644-4550.

The address is: 805 Third Avenue, New York, NY 10022.

IF MURRER DENIES THE ABOVE, PLEASE HAVE HIM PUT A STATEMENT IN WRITING.

SEND A COPY TO THIS EDITOR AT P.O. BOX 1791, APTOS, CALIFORNIA 95001. WE WILL TAKE IT FROM THERE.

Conclusions

1. The above information generates **highly disturbing conclusions.**

2. **The Administration is conning the American public.** The Administration is **playing a double game.** It creates highly publicized "spy trials" and proclamations about stopping the flow of military technology to the Soviet Union.

Why? Because the Administration **knows that over 150 Soviet weapons systems are based on U.S. technology.**

3. On the other hand, Mr. Reagan has ordered a **step-up in transfers** and the Administration is supporting **a Soviet technical vacuum operation in New York.** The Administration is checking the flow of "secrets" for public consumption while **secretly aiding** the transfer of technology behind the scenes to manufacture the previously stolen "secrets."

4. Don't waste your time writing angry letters to your Congressman. They don't know enough or claim it's all perfectly legal. The vital task ahead in 1986 is to make the information — **and more in our files** — widely known. We are lifting the copyright on this issue. Please xerox and circulate as widely as you can.

5. **We have to organize a nationwide conference on TECHNICAL TRANSFERS: IS IT TREASON?, with all the many concerned people, organizations and interests.** This means we have to forget incidental differences and **unite on this one issue,** i.e., the survival of the United States, at least until we achieve a sane policy. More next month.

Appendix I:
U.S. Weapons Technology Sold To Soviets
From San Jose Mercury News — Sept. 6, 1985

Spanish company pleads guilty to illegal exports

WASHINGTON — The Soviet Union illegally has obtained "state-of-the-art" American equipment that could help it close the gap between its weapons and highly sophisticated U.S. weaponry, and additionally highly sensitive equipment has reached Cuba, according to a Department of Commerce official and an indictment made public Thursday.

Details of the case, which involves efforts by the Soviet bloc to obtain equipment crucial to the production of highly sought computer semiconductors and integrated circuits, emerged when a Spanish company that maintains offices in Illinois agreed to pay a criminal fine of $1 million for illegally exporting high-technology equipment between 1979 and 1982.

U.S. Attorney Joseph E. diGenova issued a statement saying the violation, by Piher Semiconductores, S.A., of Barcelona, was "one of the most significant in the area of United States high-technology transfer."

Under an agreement between the Department of Justice and Piher, the company pleaded guilty to two felony counts, waived indictment by a grand jury and agreed to pay the fine. In addition, the company, which already has been barred for two and a half years from exporting U.S.-made products, will remain barred for an additional nine months.

Equipment valued at $2.4 million was shipped to the Soviet Union and Cuba, and other highly sensitive items did not get through, according to Pentagon and Department of Commerce officials familiar with the case.

Those officials described the lot as items at the top of the Soviets' list of material needed to help them move into the age of highly sophisticated, computer-dependent weapons.

"They have a major need for it in the military," said one Pentagon official, speaking on the condition he not be identified. "It would probably narrow the gap considerably in weapons systems, lending a qualitative edge to their quantitative edge."

Officials said the Soviet Union, which in the past has tried to obtain semiconductors and integrated circuits produced in the West, recently had shifted its emphasis to obtaining the equipment needed to manufacture the circuitry — the miniature wires that carry electronic data in such common gadgets as pocket calculators and digital watches, as well as in the most sophisticated space weapons.

"Such equipment is among the Soviet bloc's most highly sought American high-technology goods needed for expanding and improving the bloc's lagging microprocessor and semiconductor production capability," said Donald Creed, a Department of Commerce spokesman.

He said departmental documents confirm that $2.4 million of these goods were illegally re-exported to Cuba and Russia . . . The most sensitive, state-of-the-art semiconductor manufacturing equipment went to the Soviet Union," after first being shipped to Switzerland.

Creed said the material shipped to Cuba, and additional equipment the Cubans were unable to obtain, "would have given them the capability to produce semiconductors and integrated circuits."

"As far as we know, the plant didn't get into production," he said. "They didn't get everything they needed." However, according to the agreement accepted by Piher, Cuba already has a semiconductor manufacturing facility in Pinar del Rio.

The indictment said two senior officers of the Spanish company, Jose Puig Alabern and Francesc Sole I Planas, reached agreements with Soviet and Cuban trade organizations to obtain the equipment from U.S. manufacturers. The two are believed to be in Spain and out of reach of U.S. law enforcement officials.

Piher itself apparently does not make semiconductor manufacturing equipment, according to a spokesman for a Silicon Valley market research firm.

"To my knowledge, they don't make equipment," said Jerry Hutcheson, president of VLSI Research Inc., a San Jose company that does market research on the semiconductor industry.

The indictment states that Puig reached an agreement with Imexin, a Cuban foreign trade organization, "to provide and erect a complete integrated circuit manufacturing facility" valued at $19 million.

It said Puig and Sole, who eventually quit the company, negotiated with Technoproimport, a Soviet foreign trade organization, to sell the Soviets "two highly sophisticated U.S.-origin integrated circuit manufacturing systems."

U.S. officials and information in the indictment said U.S. officials in Spain, checking at Piher facilities to determine whether the falsely completed export license documents were being adhered to, were shown fake equipment intended to resemble that exported by Piher.

INDEX

ORDERING MORE COPIES

Gentlemen:

I would like to order additional copies of *Best Enemy Money Can Buy*.

Please send me:

_____ 1 copy @ $12.95 plus $2.00 postage and handling.

_____ Additional copies @ $10.95 each.

_____ Please send information on case lots (40 books).

I understand that these prices are postage paid bulk rates and that I only pay $2.00 postage and handling no matter what quantity I purchase. I enclose prepayment for this order.

Name _____

Address _____

City _____ State _____ Zip _____

For faster service you can use VISA or MasterCard.

Call (406) 245-6841.

Please have your card ready.

Make checks payable to:
Liberty House Press
2027 Iris Lane
Billings, Montana 59102